# 鹹蝦燦之味

*The Ham Har Chaan Cookbook*

送給我的媽媽，我的家人和 DLDL

我的同學，我的粉絲

永遠懷念我的嫲嫲
FELICIA MARIA

在2018年11月2日離開我的好爸爸
JOHN LISOLA

TO MY MOM, MY FAMILY, AND DLDL

MY STUDENTS, MY FANS

DEDICATED TO THE MEMORY OF MY GRANDMA "AVO"
FELICIA MARIA

TO MY DADDY JOHN LISOLA WHO LEFT US ON NOV 2, 2018

# 我是鹹蝦燦

鹹蝦燦是20世紀30年代中國人使用的一個貶義詞，用於形容生活在澳門的葡籍人士。第一次世界大戰後，澳門的食品供應短缺，特別是外國進口食品和雜貨，芝士是其中之一。當地的葡國人喜歡吃芝士，但又遇到物資短缺的問題，竟然找來蝦醬代替芝士。

蝦醬與芝士的香味和口感相似，一些當地葡國人開始用蝦醬塗在多士上，作為下午茶甚至早餐。這樣，慢慢形成了一種既創新又新鮮的口味。中國人認為這是奇怪的配搭，當地葡國人卻認為這是佳餚。正因這種文化衝擊讓當地人開始稱中葡人為「鹹蝦燦」。

我爸爸是葡萄牙人，而媽媽是中國人。用「鹹蝦燦」來稱呼自己的確很恰當啊！我生於香港，長於香港，很幸運我能夠在一個結合了西方和東方文化的家庭中成長，自出生以來就說兩種母語，更接觸到中西烹飪文化和技巧。我為自己的heritage感到特別和自豪。我希望透過我的故事和食譜，能夠給你一些生活中的點滴，也能勾起你一些美麗的回憶。

# My name is Ham Har Chaan

Ham Har Chaan (literally: "fermented shrimp paste") was a pejorative term used by the Chinese in the 1930s for Chinese-Portuguese Eurasian living in Macau. After WWI, there was a shortage of food in Macau, especially imported produce and grocery items, and cheese was in short supply. The local Chinese-Portuguese found a way to satisfy their cheese craving: Replace it with Chinese fermented shrimp paste.

Claiming it had the same aroma and texture as cheese, some Chinese-Portuguese started spreading the fermented shrimp paste onto toast as an afternoon snack, and even for breakfast. This slowly turned into a new way of enjoying hot toast. The Chinese didn't think much of it, while the Chinese-Portuguese considered it a delicacy. It was this cultural divide that led people to start calling Chinese-Portuguese Eurasians Ham Har Chaan. In a way, it was meant as a joke, to laugh at them.

So I use this old way of referring to Chinese-Portuguese Eurasians on social media and as the title of my first book. I consider myself lucky having grown up in a family with both Western and Asian cultures, speaking two languages since birth as my mother tongues, and being exposed to the culinary worlds of Chinese and Macanese cooking. I feel special and proud of my heritage.

I hope this book helps you understand Macanese and Portuguese cooking and provides you with delicious and nostalgic recipes that will trigger some sweet memories.

# 葡國菜 vs 澳葡菜

經常有朋友問我在葡萄牙哪裡有好吃的葡國雞，又或是葡萄牙朋友在澳門吃過葡撻後，告訴我澳門的葡撻跟葡萄牙的Pastel de Nata（葡萄牙蛋撻）不一樣。嗯，葡萄牙菜跟澳葡菜顯然是有關係，但在很多情況下它們卻可以是兩個完全獨立的派別。

葡萄牙美食源自土著農民的烹調方式，通過幾個世紀前建立的貿易路線而獲得的食材。例如，葡式咖哩來自印度，因為在16世紀開始統治了印度果阿舊城近450年。葡萄牙咖哩偏向溫和，不像印度咖哩般用上大量香料，而色彩也沒有印度咖哩般豐富。

在20世紀70年代統治莫桑比克時，葡萄牙人開始用非洲的紅椒粉（Piri Piri）。在葡萄牙有多款由Piri Piri油做成的菜式，例如Piri Piri雞、麵包、米飯、香料、糕點、香腸和海鮮。葡萄牙菜的主要食材包括洋蔥、蒜頭和蕃茄。

澳葡菜源自澳門，由中國南方，特別是廣東菜和葡萄牙菜融合而成，深受亞洲菜式的影響。許多獨特的菜餚源於葡萄牙水手的妻子將香料混合，試圖複製歐洲菜餚而成的。例如，葡國雞跟葡萄牙菜中的Estufada Galinha（炆雞）類似，澳葡版的薯蓉青菜湯因為用上芥蘭而與葡萄牙版本的不同。而澳葡菜會用上豉油和老抽，在正宗葡萄牙食譜絕對不會用上豉油作為調味料。媽媽和很多親戚仍然用中式鐵鑊來烹調澳葡菜，而非西式炒鍋或煎鍋。所以有人說澳葡菜也是fusion菜的一種，我絕對同意啊!

我可以再花2000字以上來寫一篇關於葡萄牙菜和澳門菜之差異、有關起源和食材於過去60年變化的長篇文章。我寫這本書的目的不是為了給各位上一堂歷史課，而是邀請你來到我的美食世界，一起去體驗陪我成長的菜式。告訴你，兩款葡撻（葡萄牙或澳葡式）我都喜歡。

# Portuguese Cuisine vs Macanese Cuisine

Often, people asked me where they could find a nice Macanese chicken when visiting Portugal, or my friends from Portugal tell me after visiting Macau that Pastel de Nata (Portuguese egg tarts) are not the same in Portugal. Well, Portuguese cooking and Macanese cooking are related, but they both are also independent.

The roots of Portuguese food lie in native peasant cookery, native ingredients, and ingredients obtained through trade routes established many centuries ago. Bread, rice, spices, pastries, sausages, and seafood—especially cod—remain the staples of many Portuguese meals; other key ingredients include onions, garlic, and tomatoes. Then curry came to Portuguese cooking from India during the time Portugal ruled Goa, India (from the 1500s for almost 450 years). Portuguese curry tends to be milder because the use of spices tends to be less generous and colorful as with Indian curry. Their piri piri came from Africa during the Portuguese rule of Mozambique (until the 1970s), leading to the lovable piri piri oil used in dishes like piri piri chicken.

Macau is renowned for its flavor-blending culture, and modern Macanese cuisine may be considered a type of fusion cuisine. The cooking of Macau consists of a blend of southern Chinese (especially Cantonese) and Portuguese cuisines, with significant influences from Asia. Many unique dishes came about from the use of spice blends by the wives of Portuguese sailors trying to replicate European dishes. For example, Macanese chicken is a dish close to the Portuguese estufada galinha (Chicken Stew). There are other variations: The Macanese version of caldo verde is different from the Portugal version with its use of kale. And the use of soy sauce and dark soy sauce is also common in Macanese cooking.

Common Macanese cooking techniques include grilling and roasting as well as baking, even though baking seems to be rather uncommon in authentic Chinese cooking. But Mom and a lot of my relatives use a Chinese wok to make their Macanese dishes instead of a sauté pan or skillet.

I could easily write another 2,000 words or more about the differences between Portuguese and Macanese cooking, and the origin and changes in ingredients throughout the last six decades. But the intention of my book is not to give you a history lesson, but to invite you to the world of yummy food that I was born into and grew up with. I consider myself very lucky to live in a world with both Portuguese and Macanese influences. Oh, and by the way, I love both pastel de natas and Macanese egg tarts.

這些年物資豐盛，我們可以遍嚐世界中西美食，那Sir John的食譜有何獨之處？讓我細說……回憶在上第一課Sir John給我印象是一位充滿陽光活力的小伙子，談吐有禮，面帶笑容，上課氣氛有如私人教授，每一細節都安排很緊湊，也反映了他高要求的性格。優質食材，化繁為簡的步驟，快速完成的美食，每一位同學都在愉快的氣氛享受煮食的樂趣，這是特點之一。另外推廣澳葡菜是將他和祖母成長的快樂時光，在煮食過程中勾起美好的回憶他對祖母懷念之情。我們每人都需要食物濡養身體，食物中含有親情的愛更令到食物有feel，我也被他對祖母的愛感動了！加上每一菜譜都有其小故事，增加了個中趣味，這是另一特點。至於其他樂趣，你們必需上一課親身去體會了！

Gwenny So

In the age of culinary class proliferation, what qualities does John's workshop possess to "cut through the noise" with regards to the plethora of accessible cooking lessons?

Allow me to illuminate you on the next thrilling cuisine to hit your taste buds: An exuberant Chef John unfailingly greets his students—those seasoned and new—as they trickle in, securing their aprons, sipping cups of oolong tea and catching up with one another. Once his intrepid pupils have encircled the demonstration table, the excitement begins!

Invigorated by the day's activities, he imparts the history and culture of the synergistic Sino-Latin cuisine upon his students as he prepares the tasting dishes. Likewise, sharing Macanese culture is personal for John as his cooking repertoire was passed down to him from the family matriarch: his paternal grandmother, who was, in her own right, an exceptional home cook. Moreover, John frequently reminisces on his happy childhood and closeness he shared with his grandmother when prepping family favorites. These anecdotes grant John the opportunity to share helpful tidbits, which aid in elevating his students' own culinary masterpieces.

Naturally, due to his punctilious attitude, John utilizes only hand-selected ingredients and fastidiously plans the proceedings of each culinary experience, which is only improved upon by a provided recipe compte-rendu (a bonus for those diligent notes-takers!). In spite of what sounds like a "tedious cooking lesson," John's humor and candid approach to cooking simultaneously creates a calming yet enlivening environment for even the most anxious of neophytes. Nonetheless, the authenticity of Macanese cooking manifests itself in the sapid character of each unique dish, be it Tosta de Camarao, Feijoada, or Serradura.

In all, from the rising gastronome to the culinary connoisseur, I wholeheartedly urge everyone to join one of John's culinary tutorials to taste the adventurous flavors of Macanese cuisine!

Gwenny So

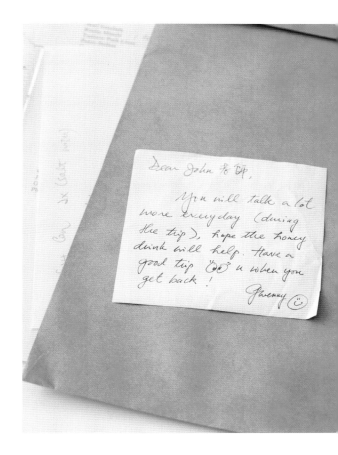

# 同學心聲

「鹹蝦燦」及「麻甩佬」兩詞都是令香港女生敬而遠之的「名詞」，但對我而言，是何等直接、沒有花巧的形容詞。愛吃的我在網絡上尋找美食之餘也會尋找美食的資料，包括食物的生長地、特性及其烹調技巧。

身為老師的我，深深明白到自己懂的事並非就能懂得教授別人。我在想，究竟怎樣的人才能真正使我理解烹調的箇中真諦？直到遇上John，一個不在食物擺盤上取巧，只在味道中追求材料的元素，更是從回憶家人及兒時的第一口味道去追求，這正是我追隨他去西葡美食之旅的原因。另外，他致力推動家鄉的葡國菜，亦使我想到很多香港人不承認自己是中國人的偉大言論，真的令人匪夷所思。

接著就是西葡美食之旅，一個真正熱血的暖男，無論從構思到實行都使我深深感受到追求理想的熱情，最重要是現代人缺少的「不忘本」，他是從嫲嫲的家庭料理菜作起步，再處理食譜。從Cooking School 小班教學，現在學生人數之多，跟著推動澳葡美食文化衝出香港，到現在寫書分享，但仍然堅持以個人味道及對澳葡菜認知為起點，不作商品為亮點，是他對澳葡菜回憶為故事的食譜，由一個愛家庭、愛國的料理老師親手執筆，這是一本愛國愛家愛味道的窩心書籍，每人都應該從中找回自己忘了的身份、國籍、家庭及味道。所以「鹹蝦燦」遠比「混血兒」來得直接，現代人太表面亦太冷漠，從John的理念中找回真我及暖流，最重要是有意思的美食!

美食不單止是「色、香、味」，更應該有「根、心、情」的存在，才是完美的美食之旅。

Michelle Ng

# 我的心底話

從小到大，我不斷有很多的ideas，所以我知道自己是一個夢想家。但我不會只是站在一角去發夢，我也是一個doer，即是我會非常實際、積極的去追夢，自己努力把夢想達成，找別人給我去完成我的夢想。這本書的誕生再不是我一個人可以做到的事情，一開始做這本書的時候，我告訴自己必定要將這篇鳴謝放在書的前部分，因為沒有你們，這個旅程是不會成事的！如果你現在正看著這篇「我的心底話」，我要向您說一聲thank you。多謝你看這本書，把它買了，帶回家，跟家人和朋友分享。

我也要在這裡多謝我的團隊，我的dream team：

Ada：多謝您對我的信任，你100%投資在這本書裡，我真的很flattered。還記得我們第一次開會的時候，你跟我說不用擔心去找贊助，不用想太多有關circulation的問題，因為你就是唯一的投資者，跟著我們的話題就是在談論「打爆機後憂鬱症」的處理方法，這個analogy引來我們大笑，這8個月的時間我有很多次都會對我這個夢想質疑，發了WhatsApp訊息給你，換來是你的dead silent。哈哈！其實你的dead silent就是對我的支持和肯定。你絕對是一個很有passion，很有heart的出版人。多謝您Ada！

Deri：你真的好細心，而我有些時候就像一隻野馬，不停的跑，有點沒有方向感的那種。而你好像已經看穿了我想的，給我提供了很多在文字上的解決方案。多謝您幫助我達成這個夢想。（Deri Reeds是我的English copyeditor）

Harry和Donald：從小到大我都是在煮家庭料理，所以對於擺盤真的know nothing。感激你們兩位提醒了我在這本書要做回自我，不要擔心用漂亮的擺盤去取悅別人。所以我們用了最直接、最original的方式去拍照。Simple is the best。

Matthew：第一次在飯局跟你見面，感覺你非常serious。但後來跟你再見面，原來你是一個非常friendly和approachable的人。跟你一起合作非常的流暢，後來你又提供了很多possible options 給我。多謝您和我一起創作了這本書！

香港人出版社團隊：多謝你們各位支持我的概念，這本書要繼續靠你們的努力給更多讀者分享。

我的助手Connie和阿欣: 多謝您們做的預備工作，還有去買食材，我們一夥人的lunch，拍攝工作真的很辛苦，如果沒有您們的協助，我真的做不來啊！

# *Words from my heart*

I am both a dreamer and doer, but putting things together and making it all look nice isn't really my forte. Writing this book has given me a good chance to know who I am, where I am standing, and where I should be heading. So I want to thank each and every one of you (and I mean you, the one reading this now) for buying this book, taking it home, cooking from it, and sharing it with your friends or relatives.

I also would like to thank the people who were instrumental in creating the book:

Ada: My boss, my boss! I was really surprised when our first meeting went so well and that you had already agreed with my idea. Thank you for giving me such a free hand to write without asking me why, and being such a good "silent partner" when I had doubts about the book. Your "play deaf" strategy reflects your trust in me. Thank you, Ada !

Deri: I salute you, Deri! You have an eye for detail, and I am sometimes like a wild horse that needs some proper guidance. I tend to talk too much and you managed to pull things together and help me polish it all in a very succinct way. We work as a very good team across the miles, and I thank you for being with me.

Harry and Donald: You guys are amazing! I am very bad with plating—this is something that everyone knows, even my mom. But you two helped me translate what I had in my mind and simply WOWed with all my dishes. Yes, simple is the best and we did it. No fuss, no bluff!

Matthew: The first time when we met at the Chinese restaurant, I was feeling a bit nervous because you acted so cool and serious. And then when we had our cooking team-building, you were so different, and funny when you added too much chicken broth to the seafood stew. LOL! Thank you for putting my castle together bits and pieces, and being so tolerant when I asked for so many changes back and forth.

The WE Press team: You are all amazing! Thank you so much for embracing my idea and pushing so hard for me.

My assistants Connie and「阿欣」: You know I wouldn't be able to make it without you two. Working on this book was really demanding, especially on photo shoot days, and you both have done so much to help me.

# 目錄 Content

第 一 部   part one

昨日 Yesterday

第二部 part two

# 今日 *Today*

第三部　part three

明日 Tomorrow

# 菜式索引 *List of Recipes*

甜品 DESSERTS

序

2016年3月4日，我在日本北海道定山溪舒舒服服enjoy我的假期。私人風呂、白色厚厚的積雪，還有每次我外遊都會帶備的一本讀物。在看著馳名作家Paulo Coelho的作品時，被書中的某句句子吸引著。便拿起了酒店提供的紙，把句子抄下來，在回程的航班上，一直想著這句說話：One day you will wake up and there won't be anymore time to do the things you've always wanted.

從沒想過自己煮的東西會得到別人的欣賞，也從沒想過自己竟然成為了一名料理導師，更加從沒想過現在為自己的書寫下這篇「序」，很多的從沒想過「never have I」。是幸運嗎？一半一半吧！說句老實話，我時常對自己說要努力把握機會，為自己的東西爭取，每天努力的工作，把家人給我的食譜重新寫過來，看參考書籍，在香港找如何做馬介休魚的方法，絞盡腦汁如何在Facebook、在Instagram的世界內生存，策劃如何做一場充滿正能量的cooking class。曾經聽過有blogger冷笑我為什麼這樣進取，我的回應是我手腳慢，所以把東西拖慢了進度，不是進取啊！

看見有些人在Facebook大鬧關注人數低，不會放錢賣廣告，用passive aggressive的方式去博粉絲同情。我每次看見這些都心中一笑，你們繼續自怨自艾吧！對我來說，沒有東西是行不通，最重要是be flexible，and never stick to one book。別人對我「眼紅」，我就會更努力，有更大的突破。

曾經想過盡一點綿力，出一本關於土生澳葡菜的食譜，作為對歷史的一個交代，又或者是名留青史。再想想，我才沒有這麼大的使命感啊！反而這一年在教料理班的過程中，我看到了很多人生百態，也在身邊遇到了很多不同的事情，有喜的，也有悲的。頓悟了？又或者是人變得更成熟吧！

快樂可以來得簡簡單單，食物可以帶給我們很多的快樂。獨樂樂，不如眾樂樂。這些日子我深深體會了這種非常滿足及實在的感覺。

這本書對我來說不是一本食譜書，而是一本充滿了「味」的書籍。有太多的人和物想跟大家分享，希望你會從第一頁開始讀起，了解每一道菜對我的意義。也希望看了一個chapter，煮了一道菜之後能夠喚起你們的一些感覺。

曾幾何時，我會問自己為何要寫書？為了名？為了利？使命感？這些都不再重要，因為我只知道 if I don't do it now，then when？這一刻我樂在其中，這已經足夠了。

2018年4月5日
香港

# *Prologue*

It was March 4, 2016. I was soaking in my private onsen villa in Jozankei, 40 minutes drive away from Sapporo Hokkaido, Japan. It was snowing and the view was just magnificent. I was reading a book by Paulo Coelho and one line made me think, and then re-think: "One day you will wake up and there won't be any more time to do the things you have always wanted." I dashed back to my bedroom and wrote those words down on a piece of paper. That piece of paper is still sitting on my writing desk now.

Never had I thought of becoming a cooking instructor, and never had I thought of having my own book, and writing this Prologue.

Macanese food is on the verge of vanishing, and less and less people know how to cook authentic Macanese dishes. When they heard about this project, friends asked me: Is it for a legacy that you are teaching and writing about Macanese cooking? Honestly, I don't think I have such a heroic vision. But I have seen a lot in a year in my cooking career, and I have changed a lot too— for good, of course. (Hahaha!)

This is not just a recipe book, and I don't see this as a cookbook itself. This is a book about you and me, about people I have met, things I have seen, and moments that I love to share with you. Please read the book from the beginning through the end, to understand my journey and appreciate the meaning of every dish I present.

In addition to my friends' questions about my cookbook, there were times that I asked myself: Why publish a book? For vanity? For fortune? Do any of these matter anymore? I'm not sure. But I do know that if I don't do it now, when? Seize the day, for I am enjoying every moment.

April 5, 2018
Hong Kong

第一部

# 昨日

沒有昨天，又怎會有今天？
小時候的成長，奠定了我對烹飪的熱愛。

# part one

# *Yesterday*

Without yesterday,
there won't be today.

# 嘩！有High Tea食!!

黑色的撥輪電話。

「8月28日是Boy Boy（我的乳名）的生日」

「4點在我們家裡有High Tea……」我記得媽媽在每通電話都是這樣說。

往後的幾個星期，媽媽和Avo（嫲嫲）就是為這個生日會忙著，澳門餅店的Macanese Pastry，還有金獅餅店的Birthday Cake（70-80年代位於中環娛樂行旁邊，即擺花街）。

如果還有時間，他們更會給我做invitation card，或到當年位於置地廣場地庫的The Hallmark Shop，買一些帶有主題式的邀請咭，然後由我親手寫上內容，寄到我aunties和uncles的家裏。And then，what is next？買新衫。

到了生日當天，媽媽、嫲嫲和工人姐姐們便會忙個非常，澳門餅店送來糕點，而媽媽更會騰空一間房間用來作為store room，將當天的物資存放在那裡，還有蘇記士多（位於羅便臣道27號）送來的支裝汽水、可樂、7-up、綠寶，還有玉泉忌廉。五顏六色的支裝汽水存放在偌大的露台裏。

跑到廚房，嫲嫲總會嚷著我和妹妹離開，但她的手已經拿著剛做好了的蝦多士給我們吃。

走到「臨時資源中心」，門給媽媽鎖了。去到客廳，偌大的飯枱已經被食物擺得密密麻麻，有西洋炸角仔Chilicote（p.35）、Cheese Toast、嫲嫲葡國雞（p.73）、咖哩、焗牛肉Estufada（p.47），而另一張較小的枱就放了甜點。如Bebinca、啫喱糖，還有我的生日蛋糕，一直吃到晚上9點。

對啊！ 這便是土生澳葡人的High Tea，又或者大家都會叫的「肥茶」＊。

小孩子的時候，我就是非常喜歡High Tea，因為High Tea就像去吃自助餐，有不同款式的東西吃。

80年代，香港出現了移民潮，我身邊很多的親人都開始離開香港，嫲嫲的身體也開始不大如前，而在家辦High Tea的次數也變少了。現在回想起，我回味當年住在羅便臣道大宅的片段，回味High Tea的時光。

＊「肥茶」Chá Gordo以豐富的土生美食招待親友。

後記：在教料理班的初期，我嘗試過設計了一個土生澳葡High Tea系列，希望通過料理班介紹澳葡High Tea文化給各位。但當時cooking school的負責人竟然將我的High Tea改為Afternoon Tea，原因是High Tea這名字太confusing，香港人不知道什麼是High Tea。我們在這觀點上argue了很久，最後我還是輸了，她把High Tea的名字改回Afternoon Tea。是我當年的名氣不夠, 粉絲人數不過20人！Well，今天我可以在這裏跟大家講：嘩！有High Tea食！

"小孩子的時候，
我就是非常喜歡High Tea"

# It's High Tea

Mom was on the phone, "It's Boy Boy's (my family nickname) birthday on August 28, and there will be a high tea at home from 4 p.m. onwards."

So in the following weeks, Mom and my avo ("grandma" in Portuguese) were busy arranging my birthday party, ordering Macanese pastries and savories from the Macau Cake Shop, and my birthday cake from the Lion Cake Shop (an old shop located near Entertainment Building, Central, Hong Kong).

And since she had the time, Avo went to The Landmark, a fancy shopping arcade in Hong Kong, and bought invitation cards from the Hallmark shop, and then asked me to write my own cards, and mail them out. What else? ... Oh yes, my new shirt and pants.

On the big day, Mom, Avo, and all the helpers were overloaded with work, and a special room was set up as a temporary storeroom for all the "yummy goodies." Oh, and there were the bottles and bottles of soft drinks delivered by So Kee grocery store—Coca Cola, Green Spot, 7-Up, etc.—all lined up neatly on our huge verandah.

My sister Karen and I ran to the kitchen, and Avo shouted at us "Out! Out of the way!" But all the while, her hand was stuffing the most delicious shrimp tosats (p.223) into our mouths. The dining table was filled with food: Feijoada (the next page), cheese toast, Avo Chicken (p.73), curry, Beef Estufada (p.47), and MaMa Salad (below). Another smaller dining table was covered with desserts: bebinca, jelly candy, and of course my birthday cake. The high tea lasted until 9 at night.

This is a typical Macanese Cha Gordo ("Fat Tea"), inspired by the English way of high tea. But the high tea I grew up with was more than just the British scones, finger sandwiches, and tea. It was a feast for family and friends, with platters and platters of local food: a huge buffet dinner with a wide array of foods for all to sample.

As years went by, more and more relatives migrated to other countries, and Avo became older and sick, so we didn't have as many high tea parties as before. But I cherish those days and all the memories from our home at Robinson Road, especially the scents of all kinds of food in the family kitchen.

# 紅腰豆炆豬手
## Feijoada

Feijoada是巴西炆燉菜式，配有各種豆類、鹹排骨和香腸。澳葡式的feijoada會用上德國鹹豬手、葡腸和圓椰菜。要在香港買到葡腸並不容易，你可以改用Spanish chorizo（西班牙香腸）。這道菜建議大家前一天準備好，吃時用白米飯來拌為最佳享受。

Feijoada is a Brazilian stew with different kinds of beans and salted ribs and sausages. The Macanese version of feijoada uses pork knuckle, linguiça, and round cabbage. As it's not easy to get linguiça in Hong Kong, you can use Spanish chorizo instead. This dish is best made a day in advance, and be sure you have enough white rice to go alongside.

| 材料（4人份量） | | Ingredient (Makes 4 servings) | |
| --- | --- | --- | --- |
| 洋蔥切碎 | 2個 | White onion, chopped | 2 |
| 橄欖油 | 1湯匙 | Olive oil | 1 tablespoon |
| 罐裝蕃茄 | 1罐（430克） | Plum tomatoes (canned) | 1 can (430g) |
| 德國鹹豬手 | 約1公斤 | Pork knuckle | about 1kg |
| 雞湯 | 500毫升 | Chicken broth | 500ml |
| 水 | 200毫升 | Water | 200ml |
| 月桂葉 | 3塊 | Bay leaves | 3 |
| 紅腰豆 | 1罐（400克） | Kidney beans | 1 can (400g) |
| 圓椰菜 | 1個 | Round cabbage | 1 |
| 葡腸或西班牙腸切片 | 1條 | Linguiça or Spanish chorizo, sliced | 1 |
| 鹽和胡椒粉 | 適量 | Salt and pepper | to taste |

做法

用鑄鐵炒鍋或不黏砂鍋，加入橄欖油把鍋燒紅，以中火將洋蔥炒至軟身，約5分鐘。加入蕃茄，邊煮邊壓碎蕃茄，直到變成糊狀，約5至8分鐘。

落豬手、葡腸、雞湯、水和月桂葉。蓋上蓋子，用細火煮，偶爾攪拌，煮約40分鐘，直到豬手軟身。

加入紅腰豆和椰菜，偶爾攪拌，繼續熬煮30分鐘，直至椰菜和紅腰豆軟身，加入鹽和胡椒粉調味。

這道菜隔天吃更惹味：將餸放涼，然後蓋上蓋子並冷藏過夜。用鑄鐵炒鍋以細火將豬手重新加熱，直至餸汁微微滾起。

How to cook

In a cast iron Dutch oven or nonstick casserole, sauté the onions in the olive oil over medium heat until soft, about 5 minutes. Add the tomatoes and cook, breaking up the tomatoes, until the tomatoes turn mushy, 5 to 8 minutes.

Stir in the pork knuckle, linguiça, broth, water, and bay leaf. Cover and cook over low heat, stirring occasionally, for about 40 minutes, until the pork knuckle is soft and tender.

Add the beans and cabbage and cook, stirring occasionally, for 30 minutes, until the cabbage is soft and the beans smear a bit when pressed. Season to taste.

The stew is best served the next day: Cool, and then cover and refrigerate overnight. Reheat in a cast iron Dutch oven over low heat until bubbling hot.

1. 如果在烹調過程中發現汁料有點乾，可加入適量雞湯。
2. 在做紅腰豆炆豬手這類炆燉菜式時，選用鑄鐵炊具總是一個明智的選擇。

1. Add more broth if the stew gets a bit dry during the cooking.
2. Cast iron cookware is always a smart choice when preparing stew dishes like feijoada.

# 西洋炸角仔
## Chilicote

西洋炸角仔Chilicotes是一種油炸的肉餡酥餅，起源於葡萄牙傳統的炸蝦餅 Shrimp Rissoles，看起來有一點像中國傳統賀年食品炸角仔。Chilicotes是我最喜歡的葡式小吃，我可以一次吃下至少10隻啊！

Chilicotes are little mince-meat-filled pastries that are deep-fried. It originated from the traditional shrimp rissoles of Portugal. The shape itself looks very much like the Chinese traditional sweet fried puff, 炸角仔, which is served during Chinese New Year.

材料（12件炸角仔份量）
餡料

| | |
|---|---|
| 橄欖油 | 1湯匙 |
| 洋蔥切粒 | 半個 |
| 蒜頭切粒 | 2瓣 |
| 咖哩粉 | 2茶匙 |
| 黃薑粉 | 1茶匙 |
| 免治牛肉 | 80克 |
| 免治豬肉（半肥瘦） | 50克 |
| 鹽和胡椒粉 | 適量 |

麵糰

| | |
|---|---|
| 水 | 200毫升 |
| 無鹽牛油 | 25克 |
| 鹽 | 1茶匙 |
| 中筋麵粉 | 200克 |
| 橄欖油或米糠油 | 炸角仔用 |

Ingredient (Makes 12 pastries)
Filling

| | |
|---|---|
| Olive oil | 1 tablespoon |
| White onion, diced | ½ |
| Garlic cloves, minced | 2 |
| Curry powder | 2 teaspoons |
| Turmeric powder | 1 teaspoon |
| Minced beef | 80g |
| Minced fatty pork | 50g |
| Salt and pepper | to taste |

Dough

| | |
|---|---|
| Water | 200ml |
| Unsalted butter | 25g |
| Salt | 1 teaspoon |
| All-purpose flour | 200g |
| Olive oil or rice bran oil | for deep frying |

做法

預備餡料：在煎鍋或平底鍋中，大火加熱橄欖油。加入洋蔥，炒至軟身。加入蒜蓉，煮1分鐘，直至香氣撲鼻。下咖哩粉和黃薑粉，煮約20秒。加入牛肉和豬肉，炒拌至肉粒分明，並且煮至全熟，約5分鐘。加入鹽和胡椒粉調味。

麵糰製法：將水、牛油和鹽混合在一個大平底鍋裡，以大火煮沸，轉細火然後下麵粉。用木勺不斷攪拌，直到麵糰形成一個球狀。

熄火，將麵糰放在枱面上，冷卻至室溫，約5分鐘。用手搓麵糰3至5分鐘，直到麵糰柔軟有彈性。

用桿麵棍將麵糰滾壓成3mm厚。用玻璃杯或曲奇餅模切出約12個小圓塊（直徑9至11厘米）

製作西洋角仔：將半湯匙餡料放在小圓塊的中心，將包滿餡料的小圓塊對摺，並用叉捏緊。

炸西洋角仔：在平底鍋或煎鍋中將6cm油以大火加熱，將一小撮麵包屑加入油中，若麵包屑在5秒內變成金黃色，這是適合炸角仔的油溫。將西洋角仔分批下鍋炸約2分鐘，轉角仔一次，炸至金黃色。撈出角仔，用廚房紙巾瀝乾油份，立即享用或放涼後享用皆可。

How to cook

For the filling: In a large skillet or saucepan, heat the olive oil over high heat. Add the onion and sauté until soft. Add the garlic and cook for 1 minute, until fragrant. Stir in the curry powder and turmeric powder and cook for about 20 seconds. Add the beef and pork and cook, stirring to break up any lumps, until cooked through, about 5 minutes. Season with salt and pepper.

For the dough: Combine the water, butter, and salt in a large saucepan. Bring to a boil over high heat. Reduce to low heat and add the flour. Stir constantly with a wooden spoon until the dough forms a ball.

Turn off the heat, place the dough on the counter, and allow to cool to room temperature, about 5 minutes. Knead the dough with your hands until soft and bouncy, 3 to 5 minutes.

Roll out the dough with a rolling pin to a 3mm thickness. Cut out about 12 small rounds (9 to 11cm in diameter) with the help of a glass or a cookie cutter.

To assemble the chilicotes: Place about ½ tablespoon of the filling in the center of one dough round. Fold the dough in half over the filling, and press the edges together to seal with a fork.

To fry the chilicotes: Heat about 6cm oil in a saucepan or frying pan over high heat until a small pinch of breadcrumbs, when added to the oil, turns brown in 5 seconds. In batches, add the chilicotes to the hot oil and fry, turning once, until golden brown, about 2 minutes. Drain on paper towels, and serve immediately or at room temperature.

"各位同學大家好！
我是你們的導師。"

# 媽媽沙律
## MaMa Salad

這不是一般的沙律：橄欖油，檸檬汁伴以少許鹽和胡椒粉。這是一個非常地道的港式沙律，你通常可以在快餐店又或是港式西餐廳找到。分享這個食譜，是因為在小時候每次的家庭聚會，我和妹妹都指定要媽媽做這款沙律給我們吃。我想跟媽媽説：Thank you so much for everything! For me, and the family.

This is not a typical salad with a dressing of olive oil and lemon juice finished with a dash of salt and pepper. It's a common "Hong Kong–style Western salad" that you might find in a local restaurant. I'm sharing the recipe with you because this is the salad that my sister and I ask Mom to make whenever we have a family gathering. And I just want to say these words to my mom: Thank you so much for everything! For me, and the family.

| 材料（5人份量） | | Ingredient (Makes enough for a family of 5) | |
|---|---|---|---|
| 青瓜切粒 | 1條 | Cucumber, diced | 1 |
| 鹽和胡椒粉 | 適量 | Salt and pepper | to taste |
| 地捫雜果（糖漿水） | 1罐（825克） | Del Monte Fruit Cocktail (in heavy syrup) | 1 can (825g) |
| 焗熟薯仔蒸熟去皮及切粒備用 | 6隻 | Baking potatoes, peeled, steamed, and diced | 6 |
| 焓蛋，去殼並切粒 | 6隻 | Hard-boiled eggs, peeled and chopped | 6 |
| 紅蘿蔔切粒 | 1隻 | Carrot, diced | 1 |
| 西芹切粒 | 1條 | Celery, diced | 1 rib |
| 卡夫奇妙醬 | 5湯匙 | Kraft mayonnaise | 5 tablespoons |

做法

將一茶匙鹽放入青瓜粒，醃15分鐘，讓青瓜出水，用乾淨毛巾或芝士布（紗布）將青瓜搾出水分，並且抹乾。

將雜果果肉和糖漿水分開。（謹記！別將糖漿水倒掉，大家可以用來做雜果賓治）

將青瓜、雜果、薯仔、蛋、紅蘿蔔和西芹倒入大碗，拌勻，逐小加入奇妙醬，最後加入鹽和胡椒粉調味。嚐嚐看！這是全世界最棒的沙律！

How to cook

Toss the cucumber with 1 teaspoon salt and let sit for 15 minutes. Transfer the cucumber to a cheesecloth or clean cloth and squeeze out all the water. Dry thoroughly.

Drain all the liquid from the fruit cocktail. (P.S.: Don't throw out the liquid! We use the syrup for making fruit punch.)

In a large bowl, combine the cucumber and fruit cocktail with the potatoes, eggs, carrot, and celery. Mix everything together and gradually add the mayonnaise. Season with salt and pepper. Enjoy it! It's the best salad in the world.

# 家傳秘技

土生澳葡菜正在失傳的其中一個原因是很少用書籍記載食譜，而由於土生澳葡菜都是以家庭菜為主，所以大部分都是一代傳一代，也漸漸地出現了很多的「家傳秘技」。

媽媽告訴我她跟爸爸結了婚一年後，我的嫲嫲才教曉她一些家傳食譜。我有一位Auntie Nonnie，有一些best kept recipe。記得每一次當她要做這些菜式的時候，一定要all clear，helpers out and door closed。而有些親戚和朋友來到香港，都會找Auntie Nonnie，希望可以嚐到她的家傳秘技。

就正如我嫲嫲的葡國雞菜式（p.73），家人跟我說這是嫲嫲的太嫲教曉她的。數一數，已經有超過90年的歷史了！

現在我開料理班，出書寫食譜，將家傳秘技公開，這樣做是逆天而行嗎？時代不同了，現在所有的東西都可以在internet上找到，而最重要的就是 it's time to change。

# Best Kept Secrets

There are not too many cookbooks covering Macanese cooking, and that's one of the reasons for the vanishing authentic Macanese cuisine. Add to that the fact that recipes are usually passed from generation to generation, without proper documentation, so important things can be lost. But most important, many families have their "best kept secret recipes," recipes that only one member of the family knows, and that are closely guarded secrets until the aunt, mother, or grandmother is ready to pass it along to a younger relative.

Auntie Nonnie, the wife of my grandma's brother, had some of the best secret recipes, with Beef Estufada being one of her signature dishes. Her recipes were so good that relatives from overseas came all the way to Hong Kong to try her dishes. Whenever she made estufada or any of her other secrets, the kitchen had to be cleared and the door was shut to keep out any curious eyes.

The Avo Chicken recipe (p.73) is another secret recipe; it's been passed from my great grandmother to my grandmother, and then to my mother. That covers 90 years!

I broke the family tradition by teaching cooking classes and including some recipes that had been secret—and now I'm also publishing a book on authentic Macanese home cooking with those recipes! Is this against my ancestors' wish? Well, well, well… The times have changed. If not now, when?

Auntie Nonnie 跟我和妹妹的合照。
她雖然非常嚴肅，卻非常痛錫我們。

This photo brings me a lot of good memories.
It was taken when I was 7 years old.
I miss Auntie Nonnie so much.
Oh, and that's Karen my sister in the photo.

# 焗牛肉
## Beef Estufada

Estufada，即是葡萄牙語「炆/燉/煨」的意思，是我最喜歡的澳葡家常菜之一。
每個葡萄牙家庭都有自己的版本。這是我的家傳食譜，是超過80多年的祖傳秘
方。傳承至今，它有著葡國菜烹調特色，同時融滙東方的烹調技術和食材。我很
喜歡這道菜，因為它是我家族歷史的一部分，而這道菜用來拌飯亦非常可口。預
先準備好這道菜，第二天翻熱吃，味道更濃郁。

Estufada, which means "stew" in Portuguese, is one of my favorite Macanese home
cooking dishes, and my comfort food. Every Portuguese family has their own version.
This is my family recipe, a secret for over 80 years—until now. It highlights the typical
Macanese way of cooking, with both Western and Eastern techniques and ingredients.
I love this dish because it's such a part of my family, but also because the sauce is so
good with rice. The stew is even better if you prepare it a day before and reheat the
next day.

| 材料（4人份量） | | Ingredient (Makes 4 servings) | |
|---|---|---|---|
| 牛腱 | 500克 | Beef shin | 500g |
| 鹽和胡椒粉 | 適量 | Salt and pepper | to taste |
| 橄欖油 | 1湯匙 | Olive oil | 1 tablespoon |
| 洋蔥切碎 | 2個 | White onions, chopped | 2 |
| 紅蘿蔔切粒 | 1個 | Carrot, diced | 1 |
| 水 | 250毫升 | Water | 250ml |
| 牛肉湯 | 500毫升 | Beef broth | 500ml |
| 雞湯 | 125毫升 | Chicken broth | 125ml |
| 老抽 | 2湯匙 | Dark soy sauce | 2 tablespoons |
| 蕃茄切碎 | 3個 | Tomatoes, chopped | 3 |
| 八角 | 2粒 | Star anise | 2 |
| 月桂葉 | 3塊 | Bay leaves | 3 |
| 冰糖 | 8克 | Rock sugar | 8g |
| 焗薯去皮及切粒備用 | 2隻 | Baking potatoes, peeled and diced | 2 |

" 這一刻我樂在其中，這已經足夠了。"

做法

將牛腱切件。用鹽和胡椒粉醃30分鐘。

用平底鍋，落油，用中火加熱。加入洋蔥，煮約5分鐘直至洋蔥軟身。再加入紅蘿蔔和牛肉，然後加水、牛肉湯、雞湯、老抽、蕃茄、八角及月桂葉。用中火煮20分鐘，偶爾攪拌。

將火調低並加入冰糖，蓋上蓋子煮10分鐘，然後加入薯仔（如果你喜歡糊狀的質地，你可以提早加薯仔），再煮30分鐘，直到牛腱軟臉。弄掉八角和月桂葉便可上碟。

How to cook

Cut the beef shin into medium-sized chunks. Season with salt and pepper and let sit for 30 minutes.

In a saucepan, heat the oil over medium heat. Add the onions and cook until soft, about 5 minutes. Add the carrot and beef and then the water, beef broth, chicken broth, dark soy sauce, tomatoes, star anise, and bay leaves. Cook over medium heat, stirring occasionally, for 20 minutes.

Reduce the heat to low and add the sugar. Cover and cook for 10 minutes, then add the potatoes. (Or you can add them earlier if you'd like a mushy kind of texture.) Cook for another 30 minutes, until the beef is tender. Remove the star anise and bay leaves and serve. Enjoy!

1. 你可以選用金錢腱代替普通牛腱，當然金錢腱的價錢較昆貴。
2. 謹記於上菜前將八角和月桂葉取出。

1. You can replace beef shin with beef shank if you prefer, although beef shank is obviously more pricy.
2. Remember to remove the star anise and bay leaves before serving. You certainly do not want to have pieces of star anise in your mouth.

# 開竅了
## Awakened

小時候的我其實不太喜歡吃東西。翻看舊照片，相中的我是thin & skinny。也正是這原因，媽媽和嫲嫲都經常擔心，絞盡腦汁的做一些美味而又極具營養的東西給我吃。所以我從小已經是一個很會吃貨的人。The only difference is 我不太喜歡吃東西。就在7歲的那個暑假，嫲嫲帶我去了菲律賓holiday，記得我是乘坐國泰航空公司客機，還參加了當年專為小朋友而舉辦的Young Discovery Club，而我們住在馬尼拉的半島酒店，早餐在酒店吃美式早餐，芒果汁，sunny side-up，不過對這些食物還是沒有太大的興趣，直至有一晚嫲嫲在馬尼拉的友人帶我們去到當地一間專門吃炸雞的餐廳，光吃了一件雞胸，我開竅了！

到現在我都不明白那塊雞有什麼魔力，只是知道它很特別，很juicy，從那一天起，我便喜歡了吃雞類的菜式。Having said that however，我還是有一類型的雞是不太喜歡吃的，就是媽媽經常都會為爸爸做的白切雞，還要配合那碗有很多乾蔥頭、生蒜和熟油做出來的醬汁。無味白炻雞，我還是要說一聲：No，thank you！

When I was a skinny seven-year-old who didn't much like to eat, Avo took me to Manila for a holiday, and to meet some of her friends and long-distance relatives.

We had room-service breakfast every day in a five-star hotel: eggs sunny side-up, freshly squeezed mango juice, and more. I'm sure all of it was delicious—but not to me. But then, one evening we went to a restaurant that served fried chicken.

That juicy, crispy, piece of chicken breast awakened my food senses! From that day onwards, my life was changed: I became that person who loves and enjoys food, especially chicken. But there's an exception: blanched chicken with shallot and garlic dressing, prepared by my mom. (Sorry, Mom!) It's a Chinese dish, loved by many people, including my dad. But... it's just not really my thing.

# 雞肉香檸薄荷清湯飯
## Chicken Rice Soup with Mint & Lemon

某天的一個晚上，突然想在最短時間內烹調美味的菜餚，但又想不出來要做什麼？更不知道如何在30分鐘內把菜做好？我的解決方案就是做一道雞湯飯。你只需要幾塊雞肉，少量米和雞湯。按照我的食譜，你就可以在不到20分鐘的時間內做出一碗美味的熱湯。

You know that evening, when you want to cook something quick and yummy, but simply have no idea what to make or how to do it in less than 30 minutes? Chicken rice soup is my solution. All you need is a few pieces of chicken, a small amount of rice, and chicken broth. Follow my recipe and you will be able to have a nice bowl of hot soup in less than 20 minutes.

| 材料（2人份量） | | Ingredient (Makes 2 servings) | |
|---|---|---|---|
| 雞腿肉（去骨和去皮）切成丁塊 | 100克 | Chicken thighs (boneless and skinless), diced | 100g |
| 鹽和黑胡椒 | 適量 | Salt and black pepper | to taste |
| 雞湯 | 500毫升 | Chicken broth | 500ml |
| 洋蔥切碎 | ½個 | White onion, chopped | ½ |
| 泰國香米 | 30克 | Uncooked Thai rice | 30g |
| 檸檬汁 | 3湯匙 | Lemon juice | 3 tablespoons |
| 紅燈籠椒，切粒 | ⅓個 | Red bell pepper, diced | ⅓ |
| 新鮮薄荷葉，切碎 | 4湯匙 | Fresh mint leaves, chopped | 4 tablespoons |

做法

用⅓茶匙鹽和黑胡椒醃雞肉。

將雞湯、洋蔥和雞肉放入一個大煎鍋裡，加入米，用中火煨至滾起。調低溫度，再煮10至15分鐘，直至飯變軟和雞肉煮熟。

把火關上，加入檸檬汁、紅燈籠椒和薄荷便完成了。

How to cook

Season the diced chicken with ⅓ teaspoon salt and black pepper to taste.

In a large saucepan, combine the broth, onion, and chicken. Add the rice and bring to a simmer over medium heat. Reduce the heat and simmer for 10 to 15 minutes, until the rice is tender and the chicken is cooked through.

Turn off the heat, add the lemon juice, bell pepper, and mint and serve.

1. 最好在享用湯之前才放薄荷葉，因為薄荷葉與熱湯只能青蜓點水，遇熱太久會變色。

2. 你可以用Risoni（也稱作orzo），一種短小的意大利粉代替米飯，煮約5分鐘，直至意粉變得有嚼勁，意大利人稱之為 al dente。

1. It's best to add the mint just before you enjoy the soup, as fragile mint leaves are not on good terms with hot liquid. LOL.

2. You can replace the rice with dried risoni (also called orzo) pasta. Cook for only about 5 minutes, just until the pasta is al dente.

# 家鄉燒雞
## Rocha Roast Chicken

2018年3月我到葡萄牙的波爾圖（Porto）旅遊時，我在一家名為Pedro dos Frangos的著名餐廳用餐（Frango葡文解作為雞），我愛上了他們的烤雞，簡單的用上蒜頭和檸檬醃製而成。餐廳經理慷慨地跟我分享了他們的食譜，我把它改良成為自家版本 —— Rocha Roast Chicken。

In March 2018, when I visited Porto, Portugal, I dined in a famous restaurant called Pedro dos Frangos (frango means chicken in Portuguese). I fell in love with their roast chicken, which is very simply marinated with garlic and lemon. The manager shared their recipe, and this is my own enhanced version—Rocha Roast Chicken.

| 材料（4人份量） | |
|---|---|
| 白酒 | 120毫升 |
| 橄欖油 | 120毫升 |
| 甜紅椒粉 | 1湯匙 |
| 辣椒粉 | 1茶匙 |
| 鹽 | 2茶匙 |
| 雞（走地雞） | 1隻（550至700克） |
| 檸檬（切半） | 1份 |
| 蒜頭去皮 | 2瓣 |
| 月桂葉 | 2塊 |

| Ingredient (Makes 4 servings) | |
|---|---|
| White wine | 120ml |
| Olive oil | 120ml |
| Paprika | 1 tablespoon |
| Cayenne pepper | 1 teaspoon |
| Salt | 2 teaspoons |
| Whole chicken (free-range) | 1 (550 to 700g) |
| Lemon (halved) | 1 |
| Garlic cloves, peeled | 2 |
| Bay leaves | 2 |

## 做法

在一個小碗裡，將白酒、橄欖油、甜紅椒粉、辣椒粉和鹽混合在一起。將雞放入大碗中，然後用混合好的醃汁塗滿整隻雞。切記要把汁料揉進雞皮內並塗勻整個雞腔。

將半個檸檬、蒜頭和月桂葉塞進雞腔內。把另一半檸檬擠出汁液並淋於雞上。把雞用保鮮紙包好並冷藏至少2小時，但不要超過4小時。

將焗爐預熱至200°C。從碗裡取出已醃好雞隻。剩餘的醃料備用。將雞隻放在架上，雞胸向上，然後將架放在中型烤盤中。將大概100毫升水倒進烤盤，將雞隻烤焗20分鐘，以鎖住肉汁。

將焗爐溫度降至160°C。將剩餘的醃料倒入烤盤。再焗40分鐘，直到雞皮略脆。將雞取出，在室溫中至少待10分鐘，然後把雞隻切成四份，即可享用。

## How to cook

In a small bowl, mix together the wine, olive oil, paprika, cayenne and salt. Place the chicken in a large bowl and rub all over with the marinade. Make sure you get under the skin and into the cavity as well.

Place 1 lemon half, the garlic cloves, and bay leaves in the cavity of the chicken. Squeeze the juice from the other lemon half all over the chicken. Cover the chicken and refrigerate for at least 2 hours, but no longer than 4 hours.

Preheat the oven to 200°C. Take the chicken out from the mixing bowl. Reserve the remaining marinade. Place the chicken on a rack, breast side up, and place the rack in a medium roasting pan. Pour about 100ml water into the pan. Roast the chicken in the oven for 20 minutes, to seal all of the juices in.

Reduce the oven temperature to 160°C. Pour the reserved marinade into the roasting pan. Roast for 40 minutes longer, until the chicken skin is slightly crispy. Let the chicken rest for at least 10 minutes before cutting it into quarters to serve.

" 準備工作感覺好像永遠做不完似的，
切蒜頭、安排食譜、試食碟、擺盤碟……
Oh My God，現場一片混亂，
我的心情是既緊張，又開心。 "

# 雞煮蜆
## Chicken with Clams

雞和蜆？每當我在料理班介紹這道菜時，我常看見學生流露出懷疑的表情。葡萄牙以其海鮮而聞名，有許多不同的方式來烹調各式各樣的海鮮。雞煮蜆就是一道老少咸宜、大人小朋友都愛吃的家常菜。任何人只要嚐一口都會對這道菜的組合感到驚訝：鮮甜飽滿的蜆肉跟啖啖雞肉就是配合得天衣無縫。　那麼我的同學怎樣評價它呢？下課時大家都豎起大拇指！

Chicken and Clams? I see a lot of surprised expressions when I introduce this dish in my cooking classes. Portugal is famous for its seafood, with many different ways of preparing all of the bounty. Chicken with clams is a common home-cooking dish that is always welcomed by people young and old. Anyone who tastes this quick dish is amazed by the combination: The freshness and juiciness of the clams blend so well with the tenderness of the chicken. It is so delicious that you don't want to waste a single drop of its sauce. And the verdict from my students after class? Thumbs up!

## 材料（4人份量）

| | |
|---|---|
| 橄欖油 | 1湯匙 |
| 中筋麵粉 | 撲粉用 |
| 鹽和胡椒粉 | 適量 |
| 雞（走地雞）切成塊 | 1隻（550至700克） |
| 紅蘿蔔切粒 | 50克 |
| 蒜頭切碎 | 2湯匙 |
| 紅辣椒片 | 1湯匙 |
| 白酒 | 120毫升 |
| 薑切碎或磨碎 | 1湯匙 |
| 雞湯（如需要） | 80毫升 |
| 新鮮蜆 | 900克 |
| 麻油 | 1茶匙 |
| 蔥切碎（裝飾用） | 適量 |

## Ingredient (Makes 4 servings)

| | |
|---|---|
| Olive oil | 1 tablespoon |
| All-purpose flour | for dredging |
| Salt and pepper | to taste |
| Whole chicken (free-range), cut into pieces | 1 (550 to 700g) |
| Carrot, diced | 50g |
| Garlic cloves | 2 tablespoons |
| Red pepper flakes | 1 tablespoon |
| Dry white wine | 120ml |
| Fresh ginger, minced or grated | 1 tablespoon |
| Chicken broth (if needed) | 80ml |
| Fresh clams | 900g |
| Sesame oil | 1 teaspoon |
| Scallions, chopped (for garnish) | small handful |

做法

在大煎鍋中以中火將油加熱。在加熱油的同時，將麵粉放在碟上，混合鹽和胡椒粉調味。將雞塊撲上麵粉，篩掉多餘的麵粉。雞肉放入鍋中煎8至10分鐘，轉一、兩次，直至變成金黃色。

從煎鍋中取出雞肉，調至中火。將所有油倒出，只剩下幾湯匙的油。加入紅蘿蔔、蒜頭和紅椒片，炒拌約5分鐘。加入白酒和薑，雞肉回鍋。

調節煎鍋的溫度，讓鍋裡食物輕輕但穩定地起泡。如果鍋裡的菜汁不足夠煨雞，則加點雞湯。蓋上蓋子，煮大概10分鐘，每2分鐘翻轉雞肉一次，直至雞肉僅僅煮熟。

將蜆放在雞肉上，再次蓋上蓋子，用中火煮3至5分鐘直至蜆殼打開，掉走未打開的蜆。拌入麻油並試味，最後用蔥花裝飾便完成了！

How to cook

Heat the oil in a large skillet over medium heat. While the oil is heating, put the flour on a plate and season with salt and pepper. Dredge the chicken pieces in the flour, shaking off excess flour. Add the chicken to the skillet and cook, turning once or twice, until nicely browned, 8 to 10 minutes.

Remove the chicken from the skillet and reduce the heat to medium. Spoon off all but a couple of tablespoons of the oil. Add the carrot, garlic, and pepper flakes and cook, stirring, for about 5 minutes. Stir in the wine and ginger, then add the chicken.

Adjust the heat so that the mixture bubbles gently but steadily. Add broth if the pan gets too dry and there is not enough liquid to braise the chicken. Cover and cook, turning the pieces every 2 minutes, until the chicken is almost cooked through, about 10 minutes.

Nestle the clams among the chicken pieces, cover again, and raise the heat to medium. Cook until the clams open, 3 to 5 minutes. Discard any clams that do not open. Stir in the sesame oil, taste and adjust the seasoning, then garnish with scallions. Enjoy!

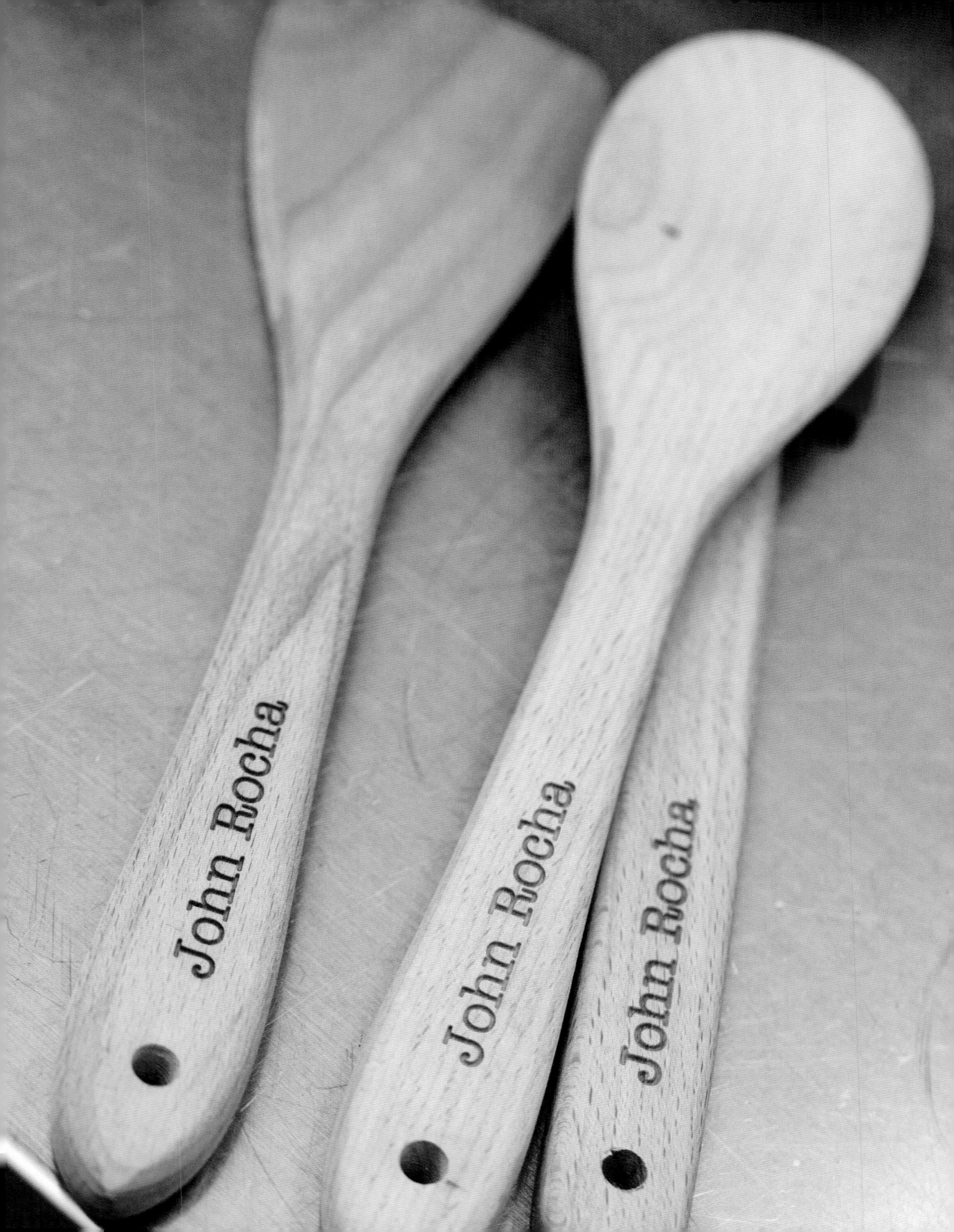
John Rocha
John Rocha
John Rocha

# 我的第一次

第一次入廚是在我讀小學二年級的時候。那一年，住在San Francisco的Auntie Anita 來香港探望嫲嫲，暫住我們的家一個月。有一天Auntie Anita跟我説要教我煮意大利粉，她便帶我到超級市場買食材，也應該是我首次認真地在supermarket買dry pasta和做意粉的sauce。

Ragu成為我首個認識到的pasta sauce，肉醬意粉便成為我首道會煮的菜式了。這個肉醬意粉並不難煮，而我就把它變為不同的菜式，用spaghetti、用lasagna和penne，這個肉醬由那時起更成為我們家中party的must have item。哈哈哈！

學懂做肉醬意粉後，我便開始跟媽媽和嫲嫲學做其他的菜式，有空的時候也會看看在家裏的cookbook。在電視上看到一些食譜，記著便把它煮了下來。嗯，就是憑記憶的煮下來。而我也自創了一些食譜，例如新派焗蛋白布甸（p.265）。

有一年的聖誕節，我便一次過把我懂得煮的菜式show hand，那些食物吃了幾天才吃完啊！

後記：那一次是我第一次見Auntie Anita，也是最後的一次。嫲嫲在1990年離開後，我們已經沒有再聯絡。但她當天在廚房教我的情景，我還是記憶猶新。

# My First Cooking Experience

I was seven years old, and Auntie Anita was visiting from San Francisco, staying with us in Hong Kong for almost a month. One afternoon she dragged me to the supermarket for shopping as she wanted to show me how to make her famous spaghetti.

We picked out the pasta (the first time I had a proper introduction in choosing a package of pasta), and of course a "ready-to-go" pasta sauce—Ragu. She showed me the Ragu brand and I remember every word she told me about how to use this magical sauce. She told me to first sauté onions until they are soft and turning brown (I consider this to be my first tip on how to sauté onions). Then add diced mushrooms and cook, always making sure the frying pan is hot. Only when the mushrooms were browned should you add the Ragu tomato sauce.

For the next couple of years, pasta with Ragu sauce was my signature dish. I was asked to cook it at every family occasion—birthday parties, Christmas gatherings, and even just friends getting together. Over time, I added twists, making it with lasagna, penne, and other pasta types. Oh yes, I even had my own version of cold pasta.

But that little success with Ragu pasta sauce took me to the next level, and I started trying other recipes from my avo and my mom, and I soon started taking notes from TV cooking shows.

P.S. That was the first time and last time I saw Auntie Anita. We lost contact after Avo passed away in 1990.

# 我的嘛嘛

我印象中的嘛嘛，是一個樣子像中國人，但一開口便是流利英式英語的老人家。嘛嘛就像一個英式的貴婦，還記得小時候我們到位於中環的西洋會所（Club Lusitano）吃飯，那裡的職員總會稱我嘛嘛為「羅夫人」。翻看舊相片，嘛嘛年輕時就是一名典型貴婦。

不過，印象最深刻的應該是嘛嘛在廚房煮東西的模樣，她對每道菜的要求，她做給我和妹妹吃的芒果布甸、炸薯仔粒（p.79）和她非常引以為榮的葡國雞。

是的，嘛嘛做的葡國雞跟別的不同，媽媽告訴我那個食譜已經超過90年歷史，原來是嘛嘛的太嘛教曉她的。嘛嘛如果還在世，今年已經接近110歲了！我諗她一定沒想過我有一天會把她的食譜公開，還開料理班做cooking instructor。她是會開心的，and she should be feeling proud of me too。

到了現在我還是非常懷念嘛嘛，是她教曉了我在小孩子時候已懂得的西方餐桌禮儀，如何用刀叉去吃雞翼，正確吃半生熟蛋的方法，當然還有很多做人道理，和對待別人的態度，還記得她躺在醫院病床的情景，還有當天的下午……。

我把葡國雞這個食譜定名為嘛嘛葡國雞「Avo Chicken」就是作為我對嘛嘛的思念和敬意。她也是我在廚藝世界裡的啟蒙老師。

嘛嘛：我非常掛念您！

# My Avo

My avo (grandma) looked like a typical Chinese woman, with a round face and black hair, but she was unlike most of the Chinese mothers and grandmothers I knew in that she spoke eloquent English, wore Western dress, and could not write even one Chinese word. She acted like a lady and I remember that she was even addressed as Lady Rocha every week when she walked into Club Lusitano, the century-old Portuguese club that was like a town hall for the Portuguese community, to play Chinese mahjong with her friends.

But what I remember most is Avo cooking in her kitchen, her very particular ways about food preparation, and the smell of the food: Potato Croutons (p.79), mango pudding, and of course her very best kept secret home recipe, Macanese chicken.

Our family's Macanese chicken is very different from the usual dish, which is yellowish and has a strong coconut flavor. Mom told me our family recipe is over 90 years old: It was passed down from my great-grandmother, to my Avo, and then to my mom.

Avo was my mentor in many ways, but especially in the kitchen. However, I am quite sure she would never have thought of her skinny grandson would turn into a cooking instructor and have such a strong passion for cooking because I never showed much interest for cooking in front of her. I miss her so much. If Avo were still with us, she would be over 110 years old!

I call our Macanese chicken family recipe Avo Chicken in memory to my avo. I see her as my teacher in cooking.

I miss you, Grandma!

# 嫲嫲葡國雞
## Avo Chicken

這道菜的原名是Macanese Chicken，因為這一切都是在澳門開始的。在葡萄牙實際上沒有這樣的菜餚，而這道菜的烹飪方式很少在傳統葡萄牙菜餚中出現。每個中葡家庭都有自己的秘方去製作香味濃郁的葡國雞。澳葡菜式經常慣用椰奶；有些家庭喜歡用上多一點的椰奶或咖哩粉令菜餚味道更濃郁。這是我的家傳食譜，從我嫲嫲（avo）的曾祖母傳給她，然後傳給我的媽媽。為了紀念親愛的嫲嫲，我將這道家傳葡國雞稱之為Avo Chicken，中文名為嫲嫲葡國雞。

The actual name of the dish is Macanese Chicken, as it all started in Macau. There's actually no such dish in Portugal and the cooking technique rarely appears in traditional Portuguese cooking. But every Portuguese-Chinese family has their own version of the delicately spiced but richly flavored chicken. The use of coconut milk is quite common in Macanese cooking; some families use slightly more coconut milk, or some add more curry powder to give a kick. This is my family's recipe, passed down from my great-grandmother to my grandma (avo), and then to my mom. I call it Avo Chicken in honor of my dear avo.

| 材料（2人份量） | | Ingredient (Makes 2 servings) | |
|---|---|---|---|
| 雞腿（連骨） | 400克 | Chicken thighs (bone-in) | 400g |
| 鹽和胡椒粉 | 適量 | Salt and pepper | to taste |
| 橄欖油 | 2湯匙 | Olive oil | 2 tablespoons |
| 雞湯或素菜湯 | 480毫升 | Chicken or vegetable broth | 480ml |
| 焗薯去皮切件 | 3隻 | Baking potatoes, peeled and cut into pieces | 3 |
| 洋蔥切碎 | 1隻 | White onion, chopped | 1 |
| 乾蔥切碎 | 2粒 | Shallots, minced | 2 |
| 蒜頭切碎 | 2瓣 | Garlic cloves, minced | 2 |
| 咖哩粉 | 2湯匙 | Curry powder | 2 tablespoons |
| 黃薑粉 | 3湯匙 | Turmeric powder | 3 tablespoons |
| 蕃茄蓉 | 240毫升 | | |

| | | | |
|---|---|---|---|
| 月桂葉 | 2至3塊 | Tomato puree | 240ml |
| 椰奶 | 300毫升 | Bay leaves | 2 to 3 |
| 焓蛋去殼切片 | 2隻 | Coconut milk | 300ml |
| 黑橄欖去核 | 12粒 | Hard-boiled eggs, peeled and sliced | 2 |
| 西班牙腸或葡萄牙香腸切片 | 1至2條 | Black olives, pitted | 12 |
| 日式麵包糠 | 1湯匙 | Chorizo or Portuguese sausage, sliced | 1 to 2 |
| | | Japanese panko | 1 tablespoon |

做法

將焗爐預熱至180°C

將雞腿切件，每件約4厘米，用鹽和胡椒粉調味。煎鍋以中火燒熱並加入1湯匙橄欖油。下雞件，煎5分鐘（在這個階段你不需要把雞肉完全煮熟）。上碟並放在一邊待用。

用中型湯鍋將上湯煨燉片刻，加入薯仔，煮20至30分鐘直至薯仔軟身。將薯仔和湯隔開，備用。

將餘下的1湯匙橄欖油放在鐵鑄烤鍋或不粘鍋上加熱。加入洋蔥、乾蔥和蒜粒，炒至軟身。加入咖哩粉和黃薑粉，煮2分鐘。加入雞肉、薯仔、湯、蕃茄蓉和月桂葉，拌勻，然後加入椰奶、鹽和胡椒粉調味。用細火煮約30分鐘，直至醬汁份量減少至一半。

將雞肉及伴菜轉到一個淺烤盤或鑲邊烤盤上，上面放上雞蛋、橄欖和西班牙香腸，再灑上麵包糠（使菜餚增添香脆的口感）。

焗10分鐘，直至表面呈金黃色且酥脆，丟走月桂葉。用白飯或麵包來拌食為佳。

How to cook

Preheat the oven to 180°C.

Chop the chicken thigh into 4cm pieces and season with salt and pepper. Heat 1 tablespoon of the olive oil in a large skillet over medium heat. Add the chicken and sauté for 5 minutes (you don't have to cook the chicken through at this stage). Transfer to a plate and set aside.

Bring the stock to a simmer in a medium saucepan. Add the potatoes and cook until soft, 20 to 30 minutes. Drain the potatoes and reserve both the stock and potatoes.

In a Dutch oven or nonstick casserole, heat the remaining 1 tablespoon olive oil. Add the onion, shallots, and garlic and sauté until soft. Add the curry and turmeric powders and cook for 2 minutes. Add the chicken, potatoes and stock, tomato puree, and bay leaves. Stir well, add the coconut milk, and season to taste with salt and pepper. Cook over low heat until the sauce is reduced by at least half, about 30 minutes.

Transfer the chicken mixture to a shallow roasting pan or rimmed baking sheet. Distribute the hard boiled eggs, olives, and chorizo over the top. Sprinkle the panko over everything (this gives it a crispy texture).

Bake for 10 minutes, until the surface is lightly browned and crispy. Find and discard the bay leaves. Serve with big trays of white rice or bread.

*Tips*

1. 你可以用一整隻雞，代替雞腿。（但我喜歡雞腿，因為它啖啖肉。）

1. You can cut up a whole chicken and use it instead of the chicken thighs. (But I love chicken thighs as they're so meaty.)

*Exercise* BOOK

# 放學後的Snacks
## After–School Snacks

我是一名70後，我的童年電視節目是《跳飛機》和後來的《430穿梭機》。

上了一整天學，回家後媽媽或嫲嫲總會在廚房預備了我喜歡吃的食物，坐在電視機前邊吃邊看，有焗蕃薯糕，也有炸薯粒。我最喜歡吃的便是Potato Croutons「炸薯粒」。

Potato Croutons是一粒一粒切到正正方方的薯仔，炸熟後用來伴免治一起吃的yummy food，外面炸得脆脆的，然後伴以免治來一齊吃，絕對是perfect！真的不知道為甚麼炸薯粒有這麼大的吸引力？我一直都喜歡吃這little snack，尤其是在剛炸好熱騰騰的時候，在薯仔粒上撒些salt and pepper，就這樣我已經心滿意足了。

記得我最喜歡的就是走到廚房，然後偷食剛炸好的Potato Croutons。為什麼每次偷食都可以成功呢？原來都是媽媽預早安排好的。她一早便發覺了，只是keep it quiet。媽媽太愛我了。

My mom made many different snacks for me when I came home from school. It was great: After a long day at school I could dig into batatada or potato croutons, while watching my favorite TV program.

By far, potato croutons was my favorite of Mom's after-school snacks. I can't find words to explain why I am so addicted to potato croutons. They are simply diced potatoes that are fried in bubbling hot oil until crispy outside and soft and moist inside. But there is something about the adorable tiny little pieces of freshly fried potatoes that make them so tempting, especially with a little dash of salt and pepper.

One of my more endearing memories was tiptoeing into the kitchen while Mom was making potato croutons and grabbing a few cubes while her back was turned, then dashing away—Mom never seemed to notice. It was years later that I learned the truth: Of course she knew what I was doing, but she just pretended not to notice and kept quiet. Well, you are so sweet Mom.

# 香脆薯仔粒
## Potato Croutons

炸薯粒是我最喜歡的小食之一，更是在我的免治菜式Minchi（p.91）中不可或缺的材料。但我必須先給各位一個溫馨提示：炸薯粒是一種會令人上癮的小食。根據我的經驗，要做出美味可口的炸薯粒，應選用焗薯或新薯（new potatoes），因為它們的質地較爽身，非常適合用來煎炸。

One of my all-time-favorite snacks, and also a M-U-S-T to include in my Minchi dish (p.91). But mind you, this is really an addictive snack that will keep dragging you back to the gym to spend hours doing your HIIT workout. Over the years, I've learned it's best to make potato croutons with baking potatoes or new potatoes as their texture is less mushy and ideal for frying.

| 材料（4人份量） | | Ingredient (Makes 4 servings) | |
|---|---|---|---|
| 薯仔去皮切粒 | 500克 | Potatoes, peeled and diced | 500g |
| 橄欖油或米糠油 | 煎炸用 | Olive oil or rice bran oil | for frying |
| 鹽和胡椒粉 | 適量 | Salt and pepper | to taste |

做法

將薯仔去皮，切成1至1.5厘米大小的薯粒。

在一個中型平底鍋中加6厘米左右的油，加熱，用木筷子浸入油中，當看見氣泡從油中升起，就是適合炸薯的油溫。

將薯粒分批落鍋煎炸，翻動薯粒一至兩次，讓薯粒均勻受熱並呈金黃色，約5分鐘。將薯粒瀝出多餘油份，放在廚房紙巾上瀝乾。

薯粒還熱哄哄的時候灑上鹽和胡椒粉調味。

How to cook

Peel the potatoes and cut into even 1 to 1.5cm cubes.

In a medium saucepan, heat about 6cm of oil until you see little bubbles coming out from the oil when you dip a pair of wooden chopstick into the oil.

In batches, add the potato cubes and fry, turning once or twice to cook evenly, until the potatoes are golden brown, about 5 minutes. Remove with a slotted spoon and drain on paper towels.

Season with salt and pepper while they are still hot.

1. 可以選擇保留薯皮，做出更有口感的炸薯粒。
2. 薯粒落鍋前，必須確保油溫夠熱。
3. 可以使用氣炸鍋。
4. 炸薯粒最好趁熱吃，混入免治肉（p.91）亦要用剛炸好的薯粒。

1. You can leave the skin on for baking potatoes or new potatoes for a more rustic texture.
2. Make sure the oil is hot enough before adding the potatoes.
3. You can use air fryer instead.
4. Potato croutons are best to serve while hot and when mixed with Minchi (p.91).

# 焗蕃薯糕
## Batatada

Batatada是一種傳統的澳葡式糕點，於下午茶和派對時用來款待客人。這款糕點有兩個做法：一款是用薯仔，另一款是用蕃薯。以蕃薯做的batatada甜而滑溜，而薯仔製成的batatada就像半硬版本的薯蓉。當我品嚐這款糕點時，我喜歡配以一杯黑咖啡。你可以把batatada放入冰箱或待在室溫冷卻後品嚐。

Batatada is a traditional Macanese potato cake that is served at high tea and parties. There are two versions: Batatada made with regular potatoes or with sweet potatoes. Sweet potato batatada is sweet and smooth, while potato batatada is like a semi-hard version of mashed potatoes. Try a slice with a strong cup of black coffee—my favorite way to enjoy it. The cake can be served cool from the fridge or at room temperature.

| 材料（4至6人份量） | | Ingredient (Makes 4 to 6 servings) | |
|---|---|---|---|
| 蕃薯去皮切塊 | 400克 | Sweet potatoes, peeled and cut into pieces | 400g |
| 無鹽牛油 | 125克 | Unsalted butter | 125g |
| 白糖粉 | 100克 | Caster sugar | 100g |
| 雞蛋輕輕打發 | 3隻 | Eggs, lightly beaten | 3 |
| 牛奶 | 135毫升 | Milk | 135ml |
| 椰奶 | 3湯匙 | Coconut milk | 3 tablespoons |
| 中筋麵粉（麵粉過篩） | 110克 | All-purpose white flour, sifted | 110g |
| 泡打粉 | 1茶匙 | Baking powder | 1 teaspoon |
| 鹽 | 少量 | Salt | pinch |

做法

將焗爐加熱至180°C，準備一個11厘米×21厘米的麵包模，掃上牛油。

將蕃薯放入蒸隔中，將水倒入長柄平底鍋中，約14至16厘米水高，放入蒸隔，蓋上蓋子並以大火將蕃薯蒸至軟身，大約20分鐘。稍微冷卻，用薯蓉器或薯蓉夾將蕃薯壓成蕃薯蓉。

用一個大碗，將牛油和糖稍為攪拌，然後加入雞蛋，輕輕拌勻。慢慢加入牛奶和椰奶，再加入蕃薯蓉。最後，用抹刀將麵粉、泡打粉和鹽與所有材料完全混合。

將麵糊倒入掃了牛油的焗模內，用180°C焗45分鐘至1小時（首30分鐘用錫紙覆蓋焗模的頂部）。要測試蕃薯糕是否熟透，你可以用刀子插入麵糊中間，如果刀身沒有黏上麵糊便代表蕃薯糕已準備好。取出焗模，放涼後切片，即可食用。

How to cook

Heat the oven to 180°C. Butter a 11cm by 21cm loaf pan.

Place the sweet potatoes in a steamer basket and place the basket in a large saucepan with 14 to 16cm water. Cover and steam over high heat until the sweet potatoes are soft, about 20 minutes. Let cool slightly and mash with a potato masher or potato ricer.

In a large bowl, beat the butter and sugar together until soft. Slowly beat in the eggs until combined. Slowly add the milk and coconut milk, followed by the mashed sweet potatoes. Finally, with a spatula, fold in the flour, baking powder, and salt until thoroughly combined.

Scrape the batter into the greased pan and bake for 45 minutes to 1 hour at 180°C, until a knife inserted in the middle comes out clean. (Cover the top of the pan with foil during the first 30 minutes.)

Let cool and slice to serve.

"要做一位稱職的料理導師is never easy，
要做一位成功的blogger或是KOL / influencer就是更難。
而我就是一位多重身份的slashie。"

# 免治的故事

近年來，爸爸一直都在生病，是肺功能衰退，所以久不久便要住進醫院。

還記得有一次爸爸住進了養和醫院，平常貪吃的他總喜歡吃養和醫院的cake，而懂得吃貨的朋友也知道養和醫院餐廳的煲仔飯也是非常popular的。當天如常的去探望爸爸，正當給他叫room service（這是我們一家人喜歡跟爸爸開的玩笑）的時候，爸爸竟然跟我說想吃免治肉配薯粒（Minchi）。

爸爸告訴我自5歲便開始吃免治肉，而他也可以每天的吃。我不是說笑，很多中葡人士都會喜歡吃Minchi。像我已經移民到加拿大超過30年的乾媽，Minchi到現在還是他們每星期必吃的菜式，Minchi對於每個中葡人都喚起了很多的回憶吧。小時候不會懂得珍惜，人大了便慢慢的學會如何去treasure一些已逝去的片段。

為了準備這部書，我在爸媽的家找來了一些他們兩人年輕時的照片，竟然給我找來了爸爸21歲生日時嫲嫲和爺爺在中環西洋會所開的生日派對。不知道為什麼，看著這照片給我很大的感觸，好像看破時間的阻隔，一幅又一幅的畫像浮現在自己的腦海裏。

爸爸對免治肉有一個很深的情義結，但對於我來說，到現在還未有一個菜式會讓我有這般大的感受。Maybe待30年後吧！不過，我倒是希望藉這本書能夠把我現在所想的、所感受的記錄下來，待我70歲時再拿出看看吧。不知道那時的世界是怎樣的呢？

謝謝你做的一切，
雙手撐起我們的家，
總是竭盡所有把最好的給我。
我是你的驕傲嗎？

時間：50年代
地點：香港西洋會所
人物：爸爸21歲的生日派對

It's Daddy's 21st birthday party at Club Lusitano, Hong Kong. And it was in the 1950s.

# The Story of Minchi

Minchi is the king of Macanese cuisine. It's a simple mix of minced beef and pork (or sometimes purely beef) cooked with soy sauce, Worcestershire sauce, and potato croutons. Every Macanese family has their own version, and every family thinks theirs is the best. The dish is so important and popular that some people have minchi a few times a week.

Take my Uncle Ronnie. the husband of my godmother, who is in love with minchi. Even after living in Canada for 30 years, they still have minchi every week, sometimes three times a week. When I visited them in Toronto, this is what they proudly served for dinner.

Daddy had his first bite of minchi when he was only five years old, and he's been relishing it all his life. Daddy is no longer as healthy as before, and has been in and out of the hospital in the past few years. At one point, he was lucky to be treated in Hong Kong's leading private hospital, which is as posh as a five-star hotel and has a renowned food service. As I was reading him the menu to help him order his lunch, he asked if, instead of the gourmet food on offer from the hospital, I could make him minchi for the next day's lunch. And dinner. Yes, I made him minchi, and made sure I included enough potato croutons.

As a comfort food for Eurasian families, I noticed that Daddy will share with me a lot of his childhood stories whenever he's enjoying the dish. So I'm wondering if there is a dish that will conjure up a lot of good old memories for me? I don't have one yet, but I think that when I turn 70, if I flip through this book, all the stories and recipes will bring up many cherished and meaningful memories.

# 澳葡式免治肉
## Minchi

材料（4至6人份量）

| | |
|---|---|
| 橄欖油 | 2湯匙 |
| 洋蔥切粒 | 半個 |
| 免治牛肉 | 500克 |
| 免治豬肉 | 100克 |
| 蒜頭切粒 | 2瓣 |
| 豉油 | 兩湯匙 |
| 噫汁 | 1湯匙 |
| 糖 | 1茶匙 |
| 炸薯仔粒（p.79） | 500克 |
| 鹽和胡椒粉 | 適量 |

Ingredient (Makes 4 to 6 servings)

| | |
|---|---|
| Olive oil | 2 tablespoons |
| White onion, minced | ½ |
| Minced beef | 500g |
| Minced pork | 100g |
| Garlic cloves, minced | 2 |
| Soy sauce | 2 tablespoons |
| Worcestershire sauce | 1 tablespoon |
| Sugar | 1 teaspoon |
| Potato Croutons (p.79) | 500g |
| Salt and pepper | to taste |

做法

用中火燒熱平底鑊，加入洋蔥，將洋蔥炒至軟身透明。加入牛肉、豬肉、蒜頭，將肉碎煮至全熟。然後加入豉油，噫汁、糖，以中火煮約10分鐘，直至餸汁餘下⅓。最後加入炸薯仔粒及鹽和胡椒粉調味，上碟。

How to cook

Heat the oil in a large skillet over medium heat. Add the onion and sauté until soft and translucent. Add the beef, pork, and garlic and cook until the meat is browned. Add the soy sauce, Worcestershire sauce, and sugar and cook over medium heat until the sauce has reduced by two-thirds, about 10 minutes. Mix in the potato croutons and season to taste with salt and pepper.

1. 如果你不喜歡豬肉，你可以全用上免治牛肉（即600克）。事實上，加入少許豬肉可令整個餸更惹味。
2. 我家人習慣在免治肉加上一隻太陽蛋，讓餸汁口感更豐富滑溜。

1. You may skip the pork if you are not a fan and use 600g of beef. But it's quite true that fatty pork will enhance the flavor of the dish.
2. My family normally puts a sunny side-up egg on top of minchi to give a moist and creamy texture.

# 澳門遊

一提及澳門，你會想起什麼？黑的？大塞車？葡撻？

我也摸不著頭腦為何中葡混血兒一定是從澳門來的？大家可否知道香港也有很多中葡混血兒，俗稱「鹹蝦燦」。而澳門對我來說就不只是關於賭場、五星級酒店、黑的……。

小時候，我經常都會「過大海」，而最不喜歡就是坐大船，還記得一坐便要坐上3小時，船名是「盧山號」。如果知道當天坐噴射船，感覺就如坐上飛機的頭等艙，飄飄然的。

10歲當年，我住在葡京酒店，記得從房間的玻璃窗向下望便是一個很大的游泳池，還有在酒店內的電子遊戲機中心。在酒店電梯大堂旁邊買鹽水鴨腎的小食亭，你們有印象嗎？晚飯的時候，我到了Uncle Jacinto的家作客，他煮了西洋冬瓜羹、咖哩雞、黃薑粉煎豬肝等等的菜式。

而每次到澳門，我的家人總會到三可餅家（現已結業）。三可餅家是一間小小的餅店，專售金錢餅、杏仁餅。在星期日，我們到教堂參與主日彌撒（對啊！用葡文的），而午餐於陸軍俱樂部，吃了非常馳名的薯蓉青菜湯（p.97）和馬介休薯球（p.143）。

# *Macau*

What do you think of when you think of Macau? Casinos? Traffic jams? Portuguese Egg Tarts? Certainly, people often think of Macau only as where Chinese-Portuguese Eurasians come from or as a gambling tourist attraction. But my impressions of Macau go beyond casinos, luxury hotels, and bad traffic.

When I was a little boy, we frequently visited Macau. I hated the ferry ride from Hong Kong, 64 kilometers across the delta, as it's such a long ride. Three hours! But whenever we took the jetfoil, I smiled. Only 50 minutes! I associated jetfoil ride as first class, while a ferry ride as coach class. Silly me!

We stayed at the Hotel Lisboa. At ten years old, I was thrilled by the hotel's huge round swimming pool and electronic game center. And the little concession store next to the lift lobby was the true highlight of our visit, especially their Chinese marinated duck liver snacks. Another highlight: dinner at Uncle Jacinto's, where he prepared Macanese-Style White Gourd Soup, Curry Chicken, Turmeric Pork Liver, and many different fish dishes.

And a trip to Macau wasn't complete without a stop at the SanHo bakery shop, which is sadly now gone. It was a tiny authentic bakery shop where we bought coin-shaped roasted almond cookies, and much more. Also clear in my memories was going to Mass on Sunday morning (performed all in Portuguese), followed by Clube Militar for lunch where I had Caldo Verde (p.97) and Bacalhau (p.143).

# 薯蓉青菜湯
## Caldo Verde

Caldo verde（字面解作「青菜湯」），被認為是葡萄牙的national dish，深受澳葡人士喜愛，更是當地人出外用膳必點的菜餚。澳葡式的薯蓉青菜湯與傳統的葡萄牙湯做法略有不同。至於如何做澳葡版本的薯蓉青菜湯？可參閱下頁內的下廚小貼士。

*Caldo verde* (literally "green soup"), considered the national dish of Portugal, is also loved by the Macanese and a must-order dish when it comes to local dining. Macanese caldo verde is slightly different from the traditional Portuguese soup—read my tips below to check for the Macanese version.

| 材料（4至6人份量） | |
|---|---|
| 初榨橄欖油 | 80毫升 |
| 蒜頭 切片 | 2瓣 |
| 洋蔥 切碎 | 1個 |
| 月桂葉 | 2塊 |
| 雞湯／素菜湯 | 500至700毫升 |
| 焗薯 (去皮，炆熟後壓成薯蓉) | 1個 |
| 葡腸切粒 | 1條 |
| 羽衣甘藍 切條 | 100克 |
| 鹽和胡椒粉 | 適量 |

| Ingredient (Makes 4 to 6 servings) | |
|---|---|
| Extra virgin olive oil | 80ml |
| Garlic cloves, sliced | 2 |
| White onion, minced | 1 |
| Bay leaves | 2 |
| Chicken or vegetable broth | 500 to 700ml |
| Baking potato, peeled, boiled, and mashed | 1 |
| Portuguese sausage, diced | 1 |
| Kale, shredded | 100g |
| Salt and pepper | to taste |

做法

將橄欖油和蒜片放入大湯鍋中，以細火煎香蒜片，然後取出蒜片並丟棄。

將洋蔥加入油中，用細火煮至軟身及半透明，再加入月桂葉。

逐少地加入高湯，然後再落薯蓉。煮沸後加入葡腸。煮約3分鐘，直至葡腸裡的油融入湯裡。加入羽衣甘藍，煮約1分鐘（烹煮羽衣甘藍不要超過兩分鐘）。取出月桂葉，加入鹽和胡椒調味即可。

How to cook

In a large soup pot, gently heat the olive oil and sliced garlic over low heat until fragrant. Remove the garlic with a slotted spoon and discard.

Add the onion to the oil and cook over low heat until soft and translucent. Add the bay leaves to the oil towards the end.

Add the broth a little at a time, followed by the mashed potato. Bring to a boil and add the sausage. Cook for about 3 minutes, until oil from the sausages releases into the soup. Add the kale and cook for about 1 minute. (Kale requires no more than a minute or two to cook.) Discard the bay leaves and season the soup with salt and pepper.

1. 澳葡式薯蓉青菜湯是使用本地芥蘭，因為它大眾化，很容易在當地市場找到。
2. 以西班牙辣肉腸代替葡腸亦可。

1. The authentic Macanese way is to use local kale（芥蘭）as it is popular and easy to find in local markets.
2. Chorizo (mild or hot) can replace the Portuguese sausage.

薯蓉青菜湯 *Caldo Verde*    99

# 黃薑粉煎豬肝
## Turmeric Pork Liver

我不是豬肝的忠實粉絲，但這是一個正宗的傳統澳葡菜，媽媽總是喜歡在一大班aunties來訪時做這個菜和大家分享。媽媽告訴我，豬膶對女士來說是上等佳餚，特別是每月的生理期。媽媽：我一定要跟我的女粉絲分享這道菜的做法啊！

I'm not a big fan of pork liver, but this is a traditional authentic Macanese dish that my mom always enjoys cooking and sharing when all the aunties are visiting. My mom told me pork liver is a delicacy for women. For PMS... So I was told. So Mom, don't you think this is a dish that I should share with all my female fans?

| 材料（4人份量） | |
| --- | --- |
| 新鮮豬肝切片 | 500克 |
| 橄欖油 | 2湯匙 |
| 蒜頭切碎 | 8瓣 |
| 乾蔥切碎 | 4粒 |
| 薑切片 | 50克 |
| 月桂葉 | 2塊 |
| 白酒 | 240毫升 |
| 檸檬汁 | 1湯匙 |
| 鹽和胡椒粉 | 適量 |
| 雞湯 | 80毫升 |
| 黃薑粉 | 2茶匙 |

| Ingredient (Makes 4 servings) | |
| --- | --- |
| Fresh pork liver, sliced | 500g |
| Olive oil | 2 tablespoons |
| Garlic cloves, chopped | 8 |
| Shallots, chopped | 4 |
| Fresh ginger, sliced | 50g |
| Bay leaves | 2 |
| White wine | 240ml |
| Lemon juice | 1 tablespoon |
| Salt and pepper | to taste |
| Chicken broth | 80ml |
| Turmeric powder | 2 teaspoons |

做法

將豬肝一直沖洗，直至沒有血水流出，水變清澈，而豬肝變成略帶深紅色，約10分鐘。

將1湯匙橄欖油倒入長柄煎鍋中以中高火加熱。加入蒜頭、乾蔥、薑片和月桂葉，炒至軟身。加入豬肝，煎至兩面金黃色，每邊半分鐘。加入白酒和檸檬汁，灑上適量鹽和胡椒粉調味。

調低至中火。用一個小碗，放入剩餘的1湯匙橄欖油、雞湯和黃薑粉並攪勻。倒入煎鍋，炒熟豬肝。如果豬肝略乾，可加少許水或湯，再輕輕翻炒，直到豬肝煮熟，將豬肝炒1分鐘左右。

How to cook

Put the liver under running tap water to rinse the blood until the water runs clear and the liver turns a slightly dark red brown, about 10 minutes.

Heat 1 tablespoon of the olive oil in a large skillet over medium-high heat. Add the garlic, shallots, ginger, and bay leaves and sauté until softened. Add the liver slices and cook, turning, until golden brown, 30 seconds on each side. Add the wine and lemon juice and season with salt and pepper.

Reduce the heat to medium. In a small bowl, mix together the remaining 1 tablespoon olive oil, the broth, and turmeric. Add to the skillet and stir fry. If the liver looks dry, add some water or more broth and lightly stir fry until the liver is cooked through, about 1 minute total.

 Tips

用什麼來配搭這道菜？大家可以試試芫荽啊！

Do you know what's a good garnish for pork liver? Chinese coriander is the answer. Try it and let me know what you think. Enjoy!

# 西洋冬瓜羹
## Macanese-Style White Gourd Soup

白葫蘆，又稱為冬瓜或中國蜜瓜，是一種瓜類蔬果，常當作蔬菜一樣食用，並用於中式菜餚。冬瓜是一種健康的食材，全年皆宜入餚。許多傳統的澳葡食譜結合了葡萄牙和中國的傳統烹飪技巧，這款西洋冬瓜羹便是一個典型例子。

White gourd, also known as wax gourd and Chinese preserving melon, is a melon-like fruit that is eaten like a vegetable and commonly used in Chinese cooking. It's a healthy food that is good to have throughout the year. Many traditional Macanese recipes combine both Portuguese and Chinese cooking traditions, and white gourd soup (crème de abobora in Portuguese) is an example because it uses lots of sautéed onions, which is Portuguese, in an otherwise Chinese soup.

| 材料（6至8人份量） | |
|---|---|
| 橄欖油 | 2湯匙 |
| 洋蔥切片 | 1個 |
| 雞湯 | 180毫升 |
| 冬瓜去皮切粒 | 500克 |
| 焗薯（去皮,焓熟後切碎） | 1個 |
| 蒜頭去衣壓成蓉 | 2瓣 |
| 鹽和胡椒粉 | 適量 |

| Ingredient (Makes 6 to 8 servings) | |
|---|---|
| Olive oil | 2 tablespoons |
| White onion, sliced | 1 |
| Chicken broth | 180ml |
| White gourd, peeled and diced | 500g |
| Baking potato, peeled, chopped, and boiled | 1 |
| Garlic cloves, peeled and smashed | 2 |
| Salt and pepper | to taste |

做法

將油放入大平底鍋或湯鍋以中高火加熱。加入洋蔥炒至軟身，但不要變成金黃色，約10分鐘。加入雞湯、冬瓜、薯仔和蒜蓉。煮10分鐘，直至冬瓜變軟。

用手提式攪拌器將所有湯料打至混合在鍋中，直至湯料變得如忌廉般滑身。如果湯太濃，你可以加一些雞湯或水來把它稀釋。

再煮5分鐘，最後加入鹽和胡椒粉調味即可。

How to cook

Heat the oil in a large saucepan or soup pot over medium-high heat. Add the onion and sauté until soft, but not browned, about 10 minutes. Add the chicken broth, followed by the white gourd, potato, and garlic. Simmer for 10 minutes, until the gourd is soft.

Use a handheld blender and mix everything right in the pot until creamy and smooth. If it's too thick and mushy, you may want to add additional broth or water to dilute.

Bring to a boil and cook for an additional 5 minutes. Season to taste with salt and pepper and serve.

我也不知道原因為何，新鮮的薄荷葉跟冬瓜湯是非常好的配搭。但謹記要在上湯前才加入薄荷葉。你不相信嗎? 來做一個實驗吧! 把薄荷葉放進一碗熱湯裡，看看5分鐘後會發生什麼事。（提示：他們變黑了）

For some I-don't-know reason, fresh mint complements white gourd soup so very well. But only add the mint leaves just before serving. If you don't believe me, do an experiment by adding mint to a hot bowl of soup, then wait for 5 minutes and see what happens. (Hint: They turn black.)

# 誰不愛Party?

葡萄牙人非常注重家庭關係，所以家庭聚會是一個非常重要的環節。而我的家庭也非常喜歡organize不同的派對。Party可以說是伴我成長。

領洗派對：葡萄牙的國教為天主教，而我也在出生後還未足三個月便領洗了。根據爸爸的回憶，家人在我領洗當天為我在家裡開了人生首個party，共有超過50人出席。而當天家人為我特別訂了一些為嬰兒領洗而做的糕點Bolo Minino，是一款以杏仁、松子仁、椰絲焗製的蛋糕。

生日派對：直到我12歲，家人每年都會為我辦生日派對，媽媽沙律（p.39）和嫲嫲葡國雞（p.73）又怎可能缺少呢？在我21歲那年，爸爸和媽媽在西洋會所為我舉辦了一個生日晚宴，邀請了本地朋友，海外朋友，還有我的鋼琴老師參加。老師還送了馬勒交響曲精裝CD給我。不說不知，我3歲已經開始學鋼琴，所以從小都喜歡聽classical music的。當晚爸爸為我安排了5道菜的晚宴：包括了炸馬介休薯球（p.143）、龍蝦湯、非洲雞、焗牛肉（p.47）和焗蛋白布甸。

還有初領聖體派對，堅振聖事派對和聖安多尼瞻禮派對。聖安多尼瞻禮的特點就是大吃水果，爸爸必會買不同種類的水果，做一個百份百的生果控。

自嫲嫲離開了之後，再加上很多aunties & uncles都相繼離開香港，我們舉辦派對的次數也越來越少。現在回想起以前派對的片段，我總覺得sweet & loving。有趣的是看著aunties & uncles可以同時用英語、廣東話和葡語交談，而我就和其他的小朋友在偌大的羅便臣道大宅內走來走去，跑到客廳把食物塞到嘴內，大人們的笑聲，唱碟機則播放著Fado音樂。此時此景，只可以成為我永久保留的一份回憶。

# We Love Parties

I am not sure if it's universal for Macanese, but my family just loves parties. The Portuguese have always valued family and we often have parties to keep everyone connected and in touch. In a way, it's part of our heritage. I have some most wonderful memories of our family parties, and of course, the food.

My baptism: Well, this was a bit early for me to actually remember, but I'm told there was a very big party after my baptism at the church, with over 50 guests in our home. Traditional pastries were made for my baptism, including bolo minino—which means "little child"— a cake baked with biscuits, almonds, pine nuts, and coconut.

My birthdays: Until I was 12, my parents threw me a birthday party every year, usually including MaMa Salad (p.39) and Avo Chicken (p.73). Then, of course, there was my 21st birthday at Club Lusitano—the club for Portuguese who live in Hong Kong. It was filled with family, friends from church, friends from overseas, my teachers—even my piano teacher, who gave me a full set of Mahler symphony CDs as a birthday gift. It was a proper sit-down five-course dinner with champagne and wine. Bacalhau Fritters (p.143) was the starter if I recall, followed by lobster bisque, African chicken, and Beef Estufada (p.47), with Molotoff for dessert.

And there were also first communion and confirmation parties, and we even celebrated feast days like St. Anthony Day where fruits were served. And on Christmas Day, we would have Daibo, the traditional Macanese stew served during winter and on special occasions.

My family hasn't had many parties since the passing of Avo and since many relatives have moved away from Hong Kong. But how I would love to travel back in time to one of these parties and watch how busy my mom and avo were, fixing all the food and crowding it on the table, with all my aunties and uncles crowding around, speaking English, Portuguese, and Chinese all at the same time, and fado music playing from the record player.

And how I'd love to see my younger self, running around in our mansion, grabbing all the delicious food in my hand and stuffing into my mouth. Those were the days.

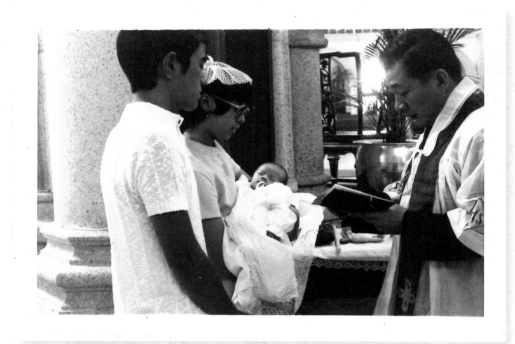

1971年9月，我在天主教總堂領洗，
圖中抱著我的是我代母 Geraldine Maria Rocha，
而旁邊的是我代父 Francis Tavares。

I was baptized in September 1971.
And that was me with my Godparents,
Geraldine (Kai Ma) and Francis (Kai Yeh).

# 西洋一品鍋
## Daibo

這是我家的傳統：在聖誕節前夕聚首一堂吃X'mas Eve晚餐，接著是聖誕日的X'mas Lunch，最後我們還會在拆禮物日一起吃大餐。在拆禮物日只有一道菜，那就是西洋一品鍋Daibo。很多中葡家庭都會將聖誕節期間慶祝時剩下來的肉類用來做西洋一品鍋：有火雞、雞肉、火腿、鵪鶉、牛肉和/或鴨肉。真正的西洋一品鍋總是用剩菜做成，這並不代表你不能用上其他食材，你可以用任何材料，例如烤雞和豬肉。這食譜是我家傳下來的版本；有時候媽媽會加上辣椒，讓這道菜更添滋味！

It's a family tradition that we all gather for dinner on Christmas Eve, followed by a lunch on Christmas Day, and finally a big dinner on Boxing Day. There's only one dish to be served on Boxing Day and that is daibo. The post-Christmas dish for every Macanese family, daibo uses all the meats left over from the festive table the day before: turkey, chicken, ham, quail, veal and/or duck. The authentic daibo is always made with leftovers, but that doesn't stop us from sometimes just making roast chicken and pork and whatever to make it. This is my family version; sometimes my mom will add chillies, which gives the dish a real kick!

| 材料（6至8人份量） | |
| --- | --- |
| 烤雞 | ½隻 |
| 已烹調豬扒 | 4塊（每塊250克） |
| 中式燒鴨 | ½隻 |
| 葡式燒豬 | ¼隻 |
| 橄欖油 | 1湯匙 |
| 洋蔥切片 | 1個 |
| 乾蔥切碎 | 2粒 |
| 蒜頭切碎 | 6瓣 |
| 雞湯 | 1至1.5公升 |

| Ingredient (Makes 6 to 8 servings) | |
| --- | --- |
| Roasted chicken | ½ |
| Pork chops, cooked | 4 (250g each) |
| Chinese roasted duck | ½ |
| Portuguese roasted pig | ¼ |
| Olive oil | 1 tablespoon |
| White onion, sliced | 1 |
| Shallots, minced | 2 |
| Garlic cloves, minced | 6 |
| Chicken broth | 1 to 1.5 liters |

| | | | |
|---|---|---|---|
| 醃製酸薑瀝出水份 | 2瓶 | Pickled ginger slices, drained | 2 jars |
| 醃製蕎頭瀝出水份 | 1瓶 | Pickled shallots, drained | 1 jar |
| 豬皮切成3厘米正方形 | 300克 | Pig skin, cut into 3cm squares | 300g |
| 法式芥末醬 | 5湯匙 | Dijon-style mustard | 5 tablespoons |
| 焗薯 | | Baking potatoes, peeled, | |
| （去皮,焙熟後切碎） | 2個 | chopped, and boiled | 2 |
| 鹽和胡椒粉 | 適量 | Salt and pepper | to taste |

## 做法

將所有肉類隨意切成一口大小的肉塊。

將油放入湯鍋或大平底鍋中用中火加熱。加入洋蔥、乾蔥和蒜頭，炒至軟身。每次加入一款已切件的肉塊，將全部煮至熱透。當鍋裡的餸汁變乾時，逐少加入雞湯。然後加入醃製過的薑片和蕎頭、豬皮及芥末，煮至醬汁變濃，約30分鐘。

最後一刻加入薯仔，煮至熱透，再用鹽和胡椒粉調味，拌以白飯。

## How to cook

Roughly chop the cooked meats into bite-size pieces.

Heat the oil in a soup pot or large saucepan over medium heat. Add the onion, shallots, and garlic and sauté until softened. Add the chopped meats one at a time, and cook until heated through. Gradually add the chicken broth when it gets too dry. Stir in the pickled ginger and pickled shallots, the pig skin, and mustard and cook until the sauce thickens, about 30 minutes.

Stir in the potatoes at the last minute and cook until heated through. Season with salt and pepper and serve with steamed white rice.

我所提供的肉類和份量只供大家參考。以我家庭為例，所用的食材和份量取決於當時剩下的是什麼菜餚或者我們現有的食物。我們通常會煮一大鍋，可供我們一周食用。我的親戚們都超愛吃這道菜！

The meats and amounts given are only suggestions. In my family, it depends on what is leftover or what we have on hand. We usually cook a huge pot of daibo that can last for the whole week. No bluffing, but my relatives just love this dish so much.

# 雪廠街16號西洋會所

小時候，我的星期天總是從主日10時英文彌撒開始，然後跟爸爸、媽媽和嫲嫲到中環吃lunch，而西洋會所便是我們經常到的用餐地方。

對於生於香港的中葡人士來説，西洋會所就如我們的「鄉親會」。而西洋會所的入會資格都是頗為嚴格，會員必須為葡籍人士，而且其爸爸也必需是葡籍人。對於葡萄牙人來説，能夠晉身這個俱樂部，於80年代或以前便是身份的象徵。我也非常lucky在21歲那年能成為西洋會所的會員。

那麼西洋會所有什麼吸引的地方？吸引著我不是在香港賽馬會內的包廂，又或是其特別高的入會門檻，而是一些令我印象難忘的美味食物。例如：混了英式芥末醬的烤芝士多士、跟芝士多士為絕配的洋蔥湯、還有焗釀蟹蓋、咖哩雞、肉丸多士和龍蝦湯。要説的真的還有很多很多。所以每次我到西洋會所我都可以吃很多美味的菜式，包括做得非常出色的青檸啫喱。對於嫲嫲，到西洋會所就一個social event，她每次都是隆重其事。

爸爸21歲當年就是在西洋會所舉行生日派對，而爸媽的結婚雞尾酒會也是在西洋會所舉行。60年後，爸爸也會如常到西洋會所吃午餐，每次他總是會吃澳葡式免治（p.91），再配一碟白飯，然後跟我説這是香港最好吃的餐廳。

# Club Lusitano

Looking back on my childhood, my usual Sunday was quite mundane, going to the 10 a.m. English mass with my avo and parents, and then lunch at either a restaurant at Central or Club Lusitano.

Oh yes: Club Lusitano. It was "the place" for Portuguese living in Hong Kong to gather. It was very difficult in the 1970s to get a membership. First, you needed to have a Portuguese family name, and your father had to be Portuguese by birth. I was "lucky" enough to become a member when I reached the age of 21.

What makes the club so special is, of course, the food. It is at Club Lusitano that I have had the best cheese toast in the world: three different kinds of cheese mixed with a mustard sauce and melted on a piece of toast. And the cheese toast goes so well with their onion soup. And then there's baked crab meat, and a chicken curry, and meat toast, and their lobster bisque. I could go on... Let's just say that going to the club is a treat for me, and I even get my favorite lime jelly. For my family, especially for my avo, going to the club is a social event.

My dad's 21st black-tie birthday party was held at Club Lusitano, and my parents had their wedding cocktail party at the club. And 60 years after his 21st birthday, Daddy still so enjoys going to the club for his lunch. I was having lunch with him the other day and he ordered a plate of Minchi (p.91) with white rice. As he ate, he kept telling me it's the best restaurant in town.

# 焗釀蟹蓋
## Baked Crab Meat

這是一道非常美味且很受歡迎的菜餚。但你知道其發源地嗎？焗釀蟹蓋乃來自巴西的街頭美食。巴西人會把餡料釀在大扇貝殼，而不是蟹殼，同時用上椰奶，而不是法式Béchamel sauce。

This is really a fancy dish that is always welcome and loved. But did you know the origin? It's actually from Brazil, where it's a gourmet street food. There, instead of a crab shell, the filling goes into a big scallop shell, and coconut milk takes the place of the béchamel.

| 材料（4人份量） | | Ingredient (Makes 4 servings) | |
|---|---|---|---|
| 橄欖油 | 2茶匙 | Olive oil | 2 teaspoons |
| 洋蔥切碎 | ⅓個 | White onion, diced | ⅓ |
| 蒜頭切碎 | ½湯匙 | Garlic, minced | ½ tablespoon |
| 紅燈籠椒切粒 | ¼個 | Red bell pepper, diced | ¼ |
| 熟蟹肉 | 300克 | Cooked crab meat | 300g |
| 帶子切粒 | 100克 | Scallops, diced | 100g |
| 黑橄欖切粒 | 2粒 | Black olives, pitted and diced | 2 |
| Béchamel sauce（見下文） | | Béchamel sauce (see below) | |
| 鹽和胡椒粉 | 適量 | Salt and pepper | to taste |
| 熟蟹殼 | 4個 | Cooked crab shells | 4 |
| 雞蛋發勻 | 1隻 | Egg, beaten | 1 |
| 日本麵包糠 | 2湯匙 | Japanese panko | 2 tablespoons |

做法

將焗爐預熱至200°C。

用中高火加熱煎鍋中的油。加入洋蔥、蒜頭和紅燈籠椒，炒至軟身。然後加入蟹肉、帶子和橄欖，炒勻。加入Béchamel sauce，用細火將海鮮和醬汁煮勻。加入鹽和胡椒調味。

將海鮮餡釀入蟹殼中，掃上蛋漿，然後在餡料上撒上麵包糠。焗5分鐘左右，直至餡面呈金黃色且略脆。

Béchamel Sauce：在平底鍋中用細火煮溶10克牛油。逐少加入10克麵粉，一直攪拌至融在一起。將125毫升牛奶分數次逐少加入，直至醬汁變得濃稠和成忌廉狀。可製成125毫升醬料。

How to cook

Preheat the oven to 200°C.

Heat the oil in a frying pan over medium-high heat. Add the onion, garlic, and bell pepper and sauté until softened. Add the crab meat, scallops, and olives and stir to combine. Add the Béchamel and cook over low heat until the seafood and sauce are well combined. Season to taste with salt and pepper.

Spoon the seafood stuffing into the crab shells, brush with the egg, and sprinkle the panko on top. Bake for about 5 minutes, until the top is brown and slightly crispy.

Béchamel Sauce: Melt 10g butter in a saucepan over low heat. Gradually add 10g flour, stirring, until incorporated. Add 125ml milk, a little at a time, until the sauce turns thick and creamy. Makes 125ml sauce.

1. 蟹殼可在街市買到。

2. 你也可以用小布甸模代替蟹殼。

1. Fresh crab shells are usually available at local wet markets.

2. Alternatively, you can use ramekins instead of the crab shells.

第二部

# 今日

每日醒來，我都是期待著。
縱使今日或有不如意的事情，我唯有接受，
然後繼續活出今日。

# part two

# Today

I wake up everyday feeling alive and looking forward to
all my challenges. There are bad days sometimes,
and I acknowledge these bad moments, and move on.

# 2017年1月11日

我人生的首場料理班始於2017年1月11日。還記得當天我安排了教授3個菜式：澳葡式洋蔥湯、黃薑粉豬肉和甜品焗椰子布甸。回想當天的我就是nervous & nervous。整天在心裏想著如何開始？要說什麼的開場白？如何烹煮？感覺自己好像是一個毫無烹飪經驗的人啊！

料理班在晚上7時開始，我4時已經到達The Cooking Alley，開始了準備的工作。不停的忙著，而準備工作感覺好像永遠做不完似的，切蒜頭、安排食譜、試食碟、擺盤碟……Oh My God，現場一片混亂，我的心情是既緊張，又開心。真的是非常複雜，猶如一句英文諺語：Butterflies in the stomach。

同學們在6時50分已到達現場。我還在廚房埋頭苦幹的在做preparation，立即告訴The Cooking Alley的負責人幫忙entertain同學，而我就把廚房門關上，合十的祈禱起來。Well，我已經有很多年沒有祈禱，我祈求上天給我一個非常順利的料理班，而在天上的嫲嫲要守護我，也希望她會以我為榮！

把廚房門打開了，我笑住的跟同學們打招呼，然後就開始了我的cooking class。同學們似乎對澳葡洋蔥湯的印象不錯啊！沒有將洋蔥和白蘿蔔的combination覺得格格不入，黃薑粉豬肉也順利過關。不說不知，我有時會擔心朋友們如何看澳葡菜，因為澳葡菜的做法和食材配搭總是不中不西。但我看見同學不斷在拍照、拍video，我知道這

場cooking class應該不錯啊！到了做甜品Bebinca的時候大家都嘩！嘩嘩！Cooking class準時於9時半完成。哈哈，是太緊張的緣故嗎？我當天累透了。

1年之後，2018年1月11日我剛剛辦了第105場料理班，學生有超過20名，而我的首場料理班只有2名學生而已。話雖如此，我仍是為每場的料理班感到緊張和開心，而preparation work仍然是做不完似的。

有一位同行前輩曾經對我說她從來不會緊張，因為她每次的開場白都是千篇一律，而她教的菜式也只是very limited，她每次在講解做蕃茄醬汁的時候，總是會用她mother-in-law的故事來做笑話。

同行前輩說教的告訴我這麼緊張是因為我的道行還未深！我聽了她的說話，只是有禮貌的笑了一笑，沒有回應。我自己心中有數啊！

到了現在，我每次上堂前都是感到既開心又緊張。就是這種感覺告訴我現在做的is the right thing。這種感覺是實在的！

# January 11, 2017

I chose three dishes to teach for my very first cooking class on January 11, 2017: Porco Bafassa, Macanese Onion Soup, and Bebinca (a multi-layered pudding). I was so excited that I didn't sleep much the night before, as I kept going over and over things in my head: *How do I start? What do I say? And how do I cook?! I sounded like an amateur. But wait, actually I AM an amateur, in that I am not professionally trained. That's the truth... would people mind? OMG!*

Class was scheduled for 7 p.m. I arrived at The Cooking Alley at 4 p.m. to start my prep work: chopping onions, mincing garlic, arranging cookware, plating dishes, organizing handouts. The place was a mess. It took me over two hours to do the prep work and it seemed never-ending. I was nervous and excited.

The clients arrived at 6:50 p.m., and I told Joyce, the owner of the cooking school, to let them chill and relax in the living room. I closed the kitchen door, put on the apron with my name on it, and said a prayer for the first time in years: I thanked God for giving me this opportunity and I hoped Avo felt proud of me. And no hiccups, please.

I smiled, opened the kitchen door, and greeted my two clients. I did a quick introduction of who I am, told a few jokes about Macanese cooking and Portuguese cooking, intending to make my clients feel relaxed (or put myself at ease,

actually), and then I started. I showed them the Portuguese onion soup first, and they were all surprised when I explained the use of Chinese radish to the soup. They tasted the soup and loved it, followed by the Bafassa. They didn't think it strange to use vinegar and turmeric powder. I was a bit skeptical about the clients' reaction on my home recipes. I sometimes thought that what I loved could be what people are scared of when it comes to food. Then they started their own cooking after tasting mine, and it went well. The dessert Bebinca, a multi-layered pudding, was another WOW! Everyone was happy and everyone kept taking photos and selfies. The class ended at 9:30 p.m., right on time.

One year later, I gave my 105th cooking classes, on January 11, 2018. This time, there were over 20 clients as compared to two for my first class. (But it still took me never-ending hours to do the prep work!)

A fellow cooking instructor, let's call her "C", told me that she's never nervous about her classes because she knows all the steps and what to say every time. But I notice that she uses almost all the same words when she teaches—she uses the same recipes over and over, she gives the same greeting at the beginning of every class, she explains her way of doing pasta sauce the same way every time, and she tells the same joke about her mother-in-law. Every time.

After hearing what C told me, I smiled politely, didn't say anything more, and continued with my prep work. Because I do still feel nervous and excited before all my classes. I like that, because those are my feelings and they mean that I know I am doing the right thing.

# 黃薑粉豬肉
## Porco Bafassa

Bafassa是一道非常地道的中葡家庭料理，就如中國家庭的紅燒豬肉。其特色就是加入白醋烹調，這樣可以平衡腩肉的油膩感。Bafassa不是葡萄牙語，而是Patois，來自澳門，一種沒落中的方言，是葡萄牙語、中文和馬來語的結合。

Bafassa is super-tender pork belly cooked with a tinge of acidity to balance the high fat content. The Portuguese word abafar, "smothered" and assar, "to roast" combine to create bafassa. It's not a Portuguese word, but from Macanese patois—a dying creole language that is a combination of Portuguese, Chinese, and Malay. My family version of bafassa skips the part of roasting, and I still love it so much.

材料（4人份量）
豬肉和醃料

| | |
|---|---|
| 豬腩肉 | 800克 |
| 黃薑粉 | ½茶匙 |
| 鹽 | 2茶匙 |
| 胡椒粉 | 1茶匙 |
| 白酒 | 100毫升 |
| 月桂葉 | 2塊 |
| 蒜頭去皮和壓扁 | 2瓣 |
| 橄欖油 | 2湯匙 |
| 洋蔥切碎 | 1個 |
| 蒜頭切碎 | 2瓣 |
| 雞湯 | 300毫升 |
| 焗薯去皮並切成四份 | 4個 |
| 水 | 200毫升 |
| 白醋 | 2至3湯匙 |
| 鹽和胡椒粉 | 適量 |

Ingredient (Makes 4 servings)
Pork and Marinade

| | |
|---|---|
| Pork belly | 800g |
| Turmeric powder | ½ teaspoon |
| Salt | 2 teaspoons |
| Pepper | 1 teaspoon |
| Dry white wine | 100ml |
| Bay leaves | 2 |
| Garlic cloves, peeled and crushed | 2 |
| Olive oil | 2 tablespoons |
| White onion, chopped | 1 |
| Garlic cloves, finely chopped | 2 |
| Chicken broth | 300ml |
| Potatoes, peeled and quartered | 4 |
| Water | 200ml |
| White vinegar | 2 to 3 tablespoons |
| Salt and pepper | to taste |

做法

於碗中將黃薑粉、鹽和胡椒粉及豬腩肉搓勻。將白酒、月桂葉和壓碎的蒜頭加入豬腩肉中。蓋上蓋子，將豬腩肉放在室溫醃製至少1小時，或放入雪櫃醃過夜。

將橄欖油倒入煎鍋或厚底砂鍋中以中火加熱。加入洋蔥和切碎的蒜頭，炒至軟身。

從醃料中取出豬腩肉（保留醃料）。將豬腩肉放落煎鍋，用大火煎約3分鐘直至變成金黃色。加入雞湯，蓋上蓋子，用中火煮10分鐘。

加入待用的醃料、薯仔、水和醋。蓋上蓋子，炆1至1.5小時，直至豬腩肉腍身。如果醬汁太少，可加水或雞湯。這個菜挺惹味，最適合用來拌飯。加入適量鹽和胡椒粉調味。

How to cook

Massage the pork belly with the turmeric, salt, and pepper and place in a bowl. Combine the wine, bay leaves, and crushed garlic and pour over the pork. Cover and marinate for at least 1 hour at room temperature, or up to overnight in the refrigerator.

Heat the olive oil in a large skillet or thick-bottomed casserole over medium heat. Add the onion and chopped garlic and sauté until soft.

Remove the pork from the marinade (reserve the marinade). Add to the skillet and sear over high heat for about 3 minutes, until browned. Add the broth, cover, and cook over medium heat for 10 minutes.

Add the reserved marinade, potatoes, water, and vinegar. Cover and simmer gently for 1 to 1½ hours, until the pork is tender. Add additional water or chicken broth if the sauce reduces too much. It should be slightly saucy and just right for serving with rice. Season to taste.

 *Tips*

1. 如果不喜歡吃豬腩肉，你可以用豬柳代替。

2. 媽媽試過用排骨來做Bafassa，但效果不及豬腩肉那麼可口。

1. If pork belly is not your cup of tea, you can use pork tenderloin.

2. My mom has tried using pork ribs, but I don't like it much.

黃薑粉豬肉 *Porco Bafassa*   133

# 澳葡式洋蔥湯
## Macanese Onion Soup

這款洋蔥湯是我其中一款正宗澳葡式家常料理食譜。媽媽告訴我這個湯的精髓是用上蘿蔔和京蔥。也可以加上少許西班牙腸或葡萄牙腸（見下廚小貼士）。這款湯的做法很簡單，不到45分鐘就能完成一碗美味熱湯。各位請慢用！

This Macanese version of onion soup is one of my authentic family recipes. Mom tells me the essence is the Chinese radish and leek. Maybe a little chorizo or Portuguese sausage as well (see the Tip on next page). It's so easy to prepare and it takes less than 45 minutes to have a nice bowl of hot soup ready. Enjoy!

| 材料（4人份量） | | Ingredient (Makes 4 servings) | |
| --- | --- | --- | --- |
| 橄欖油 | 4湯匙 | Olive oil | 4 tablespoons |
| 洋蔥切片 | 3個 | White onions, sliced | 3 |
| 紅洋蔥切片 | 1個 | Red onion, sliced | 1 |
| 京蔥切絲 | ½棵 | Leek, sliced | ½ |
| 焗薯去皮切粒 | 1隻 | Baking potato, peeled and diced | 1 |
| 蘿蔔去皮切粒 | 160克 | Chinese radish, peeled and diced | 160g |
| 雞湯或水 | 1.2公升 | Chicken broth or water | 1.2 liters |
| 鹽和胡椒粉 | 適量 | Salt and pepper | to taste |

做法

將油加入大平底鍋用細火加熱。加入兩種洋蔥，炒約10分鐘，直至洋蔥軟身但未至於金黃色。

加入京蔥、薯仔和蘿蔔，調高至中火，炒5分鐘，直至聞到京蔥的香味。

加入雞湯，以中火煮約20分鐘，直至薯仔和蘿蔔軟身。用手提式攪拌器將湯在鍋中攪拌，直至呈忌廉狀並略為變稠。依自己喜好你可以加點水來調校湯的稀釋度。加入鹽和胡椒粉調味。

How to cook

Heat the oil in a large saucepan over low heat. Add the white and red onions and sauté for about 10 minutes, until softened but not browned.

Add the leek, potatoes, and radish and turn the heat up to medium. Sauté for an additional 5 minutes, until you can smell the aroma from the leek.

Add the broth and simmer for about 20 minutes over medium heat, until the potatoes and radish are softened. Use a handheld blender to blend the soup in the pot until creamy and slightly thickened. You may want to add a little water to dilute the soup. Season to taste with salt and pepper.

 *Tips* 當湯準備好時，你可以放一至兩片西班牙腸或葡萄牙腸。我保證你會立刻愛上它。

When the soup is ready to serve, you might want to add a slice or two of chorizo or Portuguese sausage. You will love it, I guarantee.

# Annie的故事

我喜歡用料理班形式分享食譜和烹飪心得，因為料理班是互動的，溝通也是雙向的。而我也喜歡在料理班中跟各位同學說說有關食譜的history，也會傳授一些best kept secret，例如如何在家裏醃製馬介休魚。大家有講有笑，很快便消磨了3小時。

某一個星期日的上午，我如常在港島東的一間廚藝學校教授料理班，正當大家嘻嘻哈哈的時候，我開始留意到其中一位同學Annie有點異樣。Annie在料理班開始大約1小時後，無時無刻的在看著牆上的鐘。在料理班結束時，這班同學提議試食剛才煮的菜式，正當大家興高采烈的討論馬介休薯球的味道時，Annie突然嘆了一口氣，然後對我說：「John Sir，你知道嗎，我一直非常期望到這裡來上堂。因為在上課時候，我暫時可以忘記工作上的問題，拋下自己是媽媽的身份，更加不用想著下星期帶80歲高齡媽媽到醫院做check-up的問題。」對於Annie的說話，我突然變得speechless，正當在想給她一些安慰的說話時，她對著我笑，然後向我說了一聲「Thank You」。到了今日Annie仍然經常來參加我的料理班。

有時在上料理班的時候，我會站在我的導師位置，看著同學的一舉一動，有些同學在跟其他研究剛才手抄的notes，小心地在爆蒜頭，又或者互相談天。待菜式煮好後，有的已急不及待的拍下照片傳送給老公看，又或是留短訊給仔仔說：「媽媽煮了好好吃的馬介休薯球今晚帶回家。」大家離開的時候都是開開心心向我道謝。每次見到這情況，我的心都甜了，能夠用食物來帶給別人一些快樂的感覺，我已經感到非常的滿足。

# The Story of Annie

As a cooking instructor, I enjoy the interaction I have with my clients, telling them the stories of my dishes and sharing with them the little kitchen secrets, like how to make your own bacalhau at home. But there's more to teaching than that.

I remember a class I was teaching one Sunday morning. We were all enjoying ourselves and everyone loved and laughed at my bacalhau stories, and were taking notes on how to marinate a piece of cod fillet and turn it to bacalhau. But I noticed that Annie, a regular client, kept staring at the clock on the wall. She was still laughing and smiling, but I sensed something was different. At the end of the class, we were all sampling the dishes around the table and Annie suddenly sighed. She said, "Well, I have been so looking forward to coming to the class. During the class, I can be myself, forget about office work, forget about me as a mom, and stop worrying about taking my own 80-year-old mother to the hospital for a check-up." Her eyes were red and I was not sure what I should do in the situation. But then she smiled at me and said a quiet "thank you." Annie is still coming to my cooking classes and I know she's still enjoying it.

As soon as the cooking starts, I sometimes step back and look at my clients from where I stand. They are all focusing and reading my notes, making sure not to burn the chopped garlic, mixing turmeric powder with chicken broth, and chatting among themselves. Later they are smiling as they pack their dishes into their plastic containers, calling home to tell the children what's for dinner tonight, and thanking me as they leave. My role as a cooking instructor is beyond cooking; I know I can bring joy to my students, and give them some peace of mind during the few hours of cooking. And this is something that I really enjoy doing.

# 馬介休薯球
## Bacalhau Fritters

馬介休薯球是我最喜歡的菜式。馬介休「Bacalhau」是葡萄牙語中鹽醃鱈魚的意思。要處理一條馬介休魚，首先要在牛奶中浸泡24小時，以減少魚身過多鹽份並使魚肉嫩化，再在水中浸泡2小時才開始烹煮。這工序的確很花時間。

大約20年前，一些葡萄牙家庭開始自製馬介休，並且想了一個減少用鹽份量的方法。即使我們在香港，也能輕易地醃製馬介休。我把做馬介休的方法融入這食譜，但我必須再強調這是一個新派做法，而不是平常在葡萄牙或澳門旅行時吃到的傳統鹹魚。

Bacalhau fritters are, by far, my favorite food. Bacalhau is the Portuguese word for dried and salted cod. Traditionally you'd need to find a store that sells bacalhau and then soak it in milk for 24 hours to desalt and tenderize the fish, followed by 2 hours soaking in water, and only then start to cook it. It's time consuming, to say the least.

Almost 20 years ago, some Portuguese families started preparing their own bacalhau and were able to cut down on the amount of salt. Now, bacalhau is so easy to make (and cook), even in Hong Kong. This recipe includes my way of making bacalhau, but I must stress that it is the modern way, and it's not the traditional salted fish that you normally find when traveling in Portugal or Macau.

| 材料（8人份量） | | Ingredient (Makes 8 servings) | |
|---|---|---|---|
| 急凍鱈魚柳 | 450至500克 | Frozen cod fish fillets | 450 to 500g |
| 海鹽 | 1湯匙 | Sea salt | 1 tablespoon |
| 洋蔥切碎 | 3湯匙 | White onion, diced | 3 tablespoons |
| 焗薯蒸熟再壓成蓉 | ½個 | Baking potato, steamed and mashed | ½ |
| 雞蛋發勻 | ½隻 | Egg, beaten | ½ |
| 蕃茜切碎 | 少許 | Fresh parsley, minced | small handful |
| 日本麵包糠 | 1至3湯匙 | Japanese panko | 1 to 3 tablespoons |
| 橄欖油或米糠油 | 油炸用 | Olive oil or rice bran oil | for frying |

做法

製作馬介休前，先將鱈魚在室溫下解凍2小時。用鹽搓勻魚柳，用保鮮紙封好，放入雪櫃冷藏2小時。用餐巾紙印乾魚柳。將鱈魚放入蒸隔中。將水倒入平底鍋，水高5厘米，放入蒸隔和魚，蓋上蓋子然後蒸10分鐘，直至魚柳蒸熟為止。

將鱈魚印乾，將魚柳切成碎塊。小心那些細小的魚骨。

將魚碎、洋蔥、薯蓉、雞蛋和蕃茜攪勻。加入足夠的麵包糠，每次下一湯匙，直至略為黏手及麵糰沒有散開。用手或木匙羹攪拌麵糰，直至麵糰軟滑。用兩個湯匙將麵糰製成約8個小橢圓形的薯球。用保鮮紙封好，放入雪櫃至少30分鐘。

在平底鍋中倒入8厘米油，加熱，直至冒起氣泡。分批將薯球加入熱油中炸，

轉動一至兩次，直至呈金黃色。炸魚球的時間應不超過3分鐘。用罩籬將薯球取出，倒在餐巾紙上吸走多餘油份。可熱食或暖食。

How to cook

To make the bacalhau, thaw the cod at room temperature for 2 hours. Rub with the salt, cover, and refrigerate for 2 hours. Pat dry with paper towels. Place the cod in a steamer basket. Set the basket in a saucepan with 5cm of water, cover, and steam for 10 minutes, until the cod is cooked through.

Pat the cod dry and flake the fish into tiny shreds. Watch out for the tiny little bones.

Combine the shredded fish, onion, potato, egg, and parsley. Add just enough panko, a tablespoon at a time, until the texture is slightly gluey and the mixture holds together. Use your hand or wooden spoon to mix until the mixture is soft and smooth. Use two tablespoons to form the mixture into about 8 small oval fritters. Cover and rest in the refrigerator for at least 30 minutes.

Heat 8cm oil in a saucepan until bubbling hot. In batches, add the fritters to the hot oil and fry, turning once or twice, until golden brown. This should take no more than 3 minutes. Remove with a skimmer and drain on paper towels. Serve hot or warm.

馬介休薯球 *Bacalhau Fritters* 145

# 澳葡古法煮雞
## Macanese Chicken

這是另一個傳統的澳葡家常菜。每次我為朋友或家人做這菜式時,我都會準備大量白飯。這道菜的醬汁絕對令人吃上癮,要很多白飯來拌餸啊!我在煤氣烹飪中心教煮這道菜式時,我和同學用了中式鑊。為什麼用鑊?這是我嫲嫲在70年代時煮澳葡雞的方法。當然用煎鍋也是沒有問題的。

This is another authentic Macanese home cooking dish that will spoil you for sure. Whenever I prepare Macanese chicken for friends or family, I need to make sure I have cooked extra white rice. The sauce from the dish is definitely addictive, and needs plenty of rice to sop it up! I was sharing this dish in one of my cooking classes at Towngas Cooking Centre and we decided to take it a step further and use a Chinese wok instead of a large skillet. Why a wok? It's how my avo made Macanese chicken in the 1970s. If you own a Chinese wok, you have to try it and tell me what you think.

材料（4人份量）
雞肉和醃料

| | |
|---|---|
| 雞（走地雞） | 1隻（約800克） |
| 鹽 | 2湯匙 |
| 白酒 | 150毫升 |
| 橄欖油 | 3湯匙 |
| 洋蔥切碎 | 1個 |
| 乾蔥切片 | 2粒 |
| 蒜頭切碎 | 2瓣 |
| 月桂葉 | 2塊 |
| 黃薑粉 | 2湯匙 |
| 白酒 | 100毫升 |
| 雞湯 | 200毫升 |
| 鹽和胡椒粉 | 適量 |

Ingredient (Makes 4 servings)
Chicken and Marinade

| | |
|---|---|
| Whole chicken (free-range) | 1 (about 800g) |
| Salt | 2 tablespoons |
| Dry white wine | 150ml |
| Olive oil | 3 tablespoons |
| White onion, minced | 1 |
| Shallots, sliced | 2 |
| Garlic cloves, minced | 2 |
| Bay leaves | 2 |
| Turmeric powder | 2 tablespoons |
| Dry white wine | 100ml |
| Chicken broth | 200ml |
| Salt and pepper | to taste |

做法

沖洗雞隻，用紙印乾。用鹽和白酒搓勻雞隻，醃一小時。

將2湯匙橄欖油倒入煎鍋或中式鑊中以中火加熱。加入雞隻，邊煮邊翻雞隻，直到全隻雞都呈金黃色，約5分鐘。在這個階段，不需要煮熟雞隻。將雞放入碗中，待用。

將剩餘的1湯匙橄欖油落入中式鑊、鑄鐵鍋或不黏砂鍋中用中火加熱。加入洋蔥和乾蔥，炒5分鐘直至軟身。加入蒜頭和月桂葉，繼續炒約5分鐘。加入黃薑粉，立即調至細火，炒不超過30秒。加入白酒和雞湯。

雞隻回鑊，蓋上蓋子，用細火煮約30分鐘，直到雞隻徹底煮熟。將雞隻切成四份或斬件，用鹽和胡椒粉調味，即可上碟。

How to cook

Rinse the chicken and pat dry with paper towels. Rub the chicken with the salt and white wine and let marinate for an hour.

Heat 2 tablespoons of the olive oil in a large skillet or Chinese wok over medium heat. Add the chicken and cook, turning, until all sides are golden brown, about 5 minutes. The chicken does not need to be cooked through at this stage. Put the chicken in a bowl and set aside.

Heat the remaining 1 tablespoon olive oil over medium heat in a wok, Dutch oven, or nonstick casserole. Add the onion and shallots and sauté until soft, about 5 minutes. Add the garlic and bay leaves and continue to sauté for about 5 minutes. Add the turmeric, immediately turn the heat to low, and stir fry for 30 seconds or less. Add the wine and broth.

Add the chicken, cover the pot, and cook over low heat for about 30 minutes, until the chicken is thoroughly cooked. Chop or quartered the chicken, season with salt and pepper, and serve.

在烹調過程中，如果醬汁有點乾，就加多點雞湯，特別是用中式鑊烹調這道菜。

Add more broth if the sauce gets a bit dry during the cooking, especially if you are using a Chinese wok.

One day you will wake up and there won't be
anymore time to do the things you've always wanted.

# 我在少林寺的日子

如果你參加過我的cooking class，你總會發覺在我身邊會有1位助手，在一些大型的cooking class，我還會有2位助手。

回想在成為料理班導師的最初階段，我就是在The Cooking Alley渡過了猶如在少林寺習武的片段。所有的東西大部分都是我一個人準備，食材的preparation、shopping，還有分配試食的部分，可以說是集多功能於一身。而The Cooking Alley也是一個給我研究食譜及testing的地方，很多菜式如Batatada（焗蕃薯糕，p.83）及海鮮鍋都是在The Cooking Alley試做出來。這些日子雖然tedious，但也有很多難忘及快樂的片段。例如在八號風球下醃燒豬，重複焗烤葡萄牙半熟蛋糕，前前後後共做了超過10次才成功啊！

一場料理班的準備工作是非常重要的，同學們大部分都是會準時到達現場，卻很少有機會看過料理班背後的工作，就如煮一個葡萄牙海鮮鍋，每位同學需要1個切碎洋蔥、3粒蒜蓉。如果1堂有24位同學出席，我們便要準備切24個洋蔥，還有導師的portion，再剁72粒蒜頭……這些工序實在不容易啊！

很多時我在想為什麼葡萄牙菜的食材總是離不開令人催淚的洋蔥和充滿濃烈重口味的蒜頭呢？有一次我在Cooking Alley分享澳葡一品鍋（Daibo，p.113）的菜式，那一天的preparation我一個人共做了超過4小時，還未到上堂我已經倦透了。

現在非常幸運地多了助手幫忙，而我的料理班也由以往1間cooking school增加到4間，但我仍是非常緊張preparation work，很多時會親自make sure things are all right。

料理班是一個teamwork，沒有一班同事的幫忙，料理班是不行的，我也想在這裡特別多謝我的兩位助手Connie和阿欣，還有每間cooking school的同事。妳們才是最棒的！！

真的要感激The Cooking Alley 的 Joyce！

Cheers to Joyce from The Cooking Alley.

# In the Beginning

If you have attended one of my classes, you know that I normally have an assistant helping me with things, and for larger classes, I'll often have two helpers.

But looking back to when I started, I was all by myself. I started at The Cooking Alley, where I had to put serious thought into the planning and organization of my classes, testing and trying out a lot of dishes, like Batatada (a sweet potato cake, p.83) and the seafood stew Caldeirada.

There were some hilarious moments in those first classes, but more important, it's where I trained myself to become a cooking instructor and sharpened my cooking skills.

Preparation is a critical part to a successful cooking class, but it's behind-the-scenes work that clients seldom see. Baskets of onions need to be chopped or diced, cod fish needs to be steamed and marinated, and a dozen more errands and tasks need to be taken care of. The Cooking Alley is a cozy boutique cooking school, and when I started I did all that myself. So if clients arrived earlier than expected, they saw me working like a busy bee. This was the aspect that I hated about teaching cooking. Why the never-ending chopping work? I shouted in my mind while putting on a happy and positive face for the early clients. LOL. (Really, why does Portuguese cooking require so much chopped onions and garlic?) There was one class that required four hours of prep work (for the Daibo, p.113)! By the time clients arrived, I was totally exhausted.

This lasted for more than eight months, until I started taking on more classes and expanded my roster to teach at four cooking schools. But looking back, I realize that it was during those early teaching days that I learned a lot—it was such intensive training and made me become more efficient and faster in my prep work.

Nowadays, my classes are all about teamwork, and I can never achieve much without the help of other talented people. I am thankful to my two assistants, Connie and 「阿欣」, and also to all the other helpers at the cooking schools. You guys rock!

# 海鮮鍋
## Caldeirada

這是葡萄牙式和加利西亞式炆魚煲，有薯仔、各式各樣的魚配以檸檬和新鮮香草來調味。它是葡萄牙的常見菜式，通常被稱為「漁夫煲」。我將這個傳統食譜稍為變一變，剔走薯仔，仍然保留了Caldeirada的精髓。我從街市中挑選許多不同的海鮮加入這個菜式，提升這個菜餚的水平。

This Portuguese and Galician fish stew of potatoes and a wide variety of fish is flavored with lemon and fresh herbs. Often called fishermen's stew, it's a common dish in Portugal. I have modified the traditional recipe slightly by leaving out the potatoes, but still keep the essence of caldeirada. And I take the stew to the next level by adding lots of different seafood that I pick from the wet market.

| 材料（4人份量） | | Ingredient (Makes 4 servings) | |
|---|---|---|---|
| 橄欖油 | 3湯匙 | Olive oil | 3 tablespoons |
| 洋蔥切碎 | 1個 | White onion, chopped | 1 |
| 蒜頭切碎 | 2瓣 | Garlic cloves, minced | 2 |
| 罐裝蕃茄連汁 | 1罐（400克） | Plum tomatoes (canned) | 1 can (400g) including liquid |
| 雞湯 | 720毫升 | Chicken broth | 720ml |
| 白酒 | 240毫升 | Dry white wine | 240ml |
| 紅辣椒片 | 2茶匙 | Red pepper flakes | 2 teaspoons |
| 月桂葉 | 2塊 | Bay leaves | 2 |
| 鱈魚柳 | 200克 | Cod fish fillets | 200g |
| 龍蝦尾 | 4隻（每隻90g） | Lobster tails | 4 (90g each) |
| 青口 | 50克 | Blue mussels | 50g |
| 蜆 | 50克 | Clams | 50g |
| 芫茜切碎 | 少許 | Fresh cilantro, chopped | a handful |
| 蕃茜切碎 | 少許 | Fresh parsley, chopped | a handful |
| 檸檬汁 | 2茶匙 | Lemon juice | 2 teaspoons |
| 鹽和胡椒粉 | 適量 | Salt and pepper | to taste |

做法

用中火加熱中型平底鍋內的油。加入洋蔥，炒5分鐘，直至軟身。加入蒜頭，再煮一分鐘。

加入蕃茄及其汁液、雞湯、白酒、紅辣椒片和月桂葉。用中火煮10分鐘，然後加入鱈魚、龍蝦尾、青口和蜆。蓋上蓋子煮約5分鐘，直到蜆和青口打開。弄走月桂葉和仍然未打開的蜆和青口。

熄火，灑上芫茜、蕃茜和檸檬汁。用鹽和胡椒粉調味，趁熱呼呼時立即享用！

How to cook

Heat the oil in a medium saucepan over medium heat. Add the onion and sauté for 5 minutes, until softened. Add the garlic and cook for another minute.

Add the tomatoes and their liquid, the broth, wine, pepper flakes, and bay leaves. Simmer for 10 minutes over medium heat, then add the cod, lobster tails, mussels, and clams. Cover and cook for about 5 minutes, until the clams and mussels open. Discard the bay leaves and any clams and mussels that are still closed.

Turn off the heat and sprinkle with the cilantro, parsley, and lemon juice. Season with salt and pepper and serve hot. Enjoy!

# 豬柳硬碰豬扒

大家都是吃貨一族，相信都知道如何分辨豬柳和豬扒，又或是其不同的煮法。

我記得是一間料理班學校的第二班，我正準備好有關preparation的工作。在前往cooking school的途中，我發了一個WhatsApp訊息給我料理學校，查詢有關食材的狀況，提醒他們謹記把豬柳放在室溫待涼，我收到cooking school的回覆，看到那張照片，我頓時呆了！

在枱上整整放了12塊法式豬扒，非常漂亮的。但我要求的12塊豬柳卻不知所蹤。我馬上致電cooking school查個究竟，而獲得的回覆是用豬扒代替今晚我要用的豬柳。而負責的同事更告訴我可以將原有45分鐘的烹調時間減為15分鐘，以ensure肉質不會變得tough。

真的是天啊！在電話跟我通話的是在外國修畢廚藝學院的烹飪達人，而我只是一個「門外漢」。對於烹飪達人提出的要求，我真的不知道如何回應。改食譜？

我不打算跟他理論，所以自己便跑到supermarket先買來豬柳。回到cooking school，烹飪達人看見我手上拿著豬柳，冷笑的跟我説：「豬扒可以做出千變萬化的菜式，難道你不知道嗎？」我回答他：「但是今晚我是煮豬柳的菜式，客人是因為豬柳菜式而報名啊！」這是我的料理班，and I will do it my way。

料理班當天晚上很順利便完成了，而同學們也不知道在料理班前發生的小風波，那一次也是我在那所學校的最後一堂。我再沒有在那所學校再開班。有後悔嗎？當然沒有啊！

# Pork Tenderloin vs Pork Chop

I am quite sure most of you reading my book know the difference between a pork tenderloin and a pork chop. I wish I could say the same about the staff of some cooking schools.

I was looking forward to my second session at a new cooking school. Prior to the class, I had told the school staff my requirements for the class and given them a shopping list. As I was on my way, I sent a message to the school, reminding them to let the pork tenderloin "cool off" at room temperature for my seasoning. When I saw the photo they sent to me, I was in shock!

There were 12 thick French-cut pork chops lined up on the table. No pork tenderloin was in sight. I immediately called to complain, and was told that I could use the pork chops to replace the pork tenderloin, all I needed to do was reduce the cooking time from 45 minutes to 15 minutes should I worry about the tenderness of the meat.

I am not a chef, and I am not professionally trained. On the phone was someone trained in a reputable cooking academy in Europe, and I was totally speechless. I knew for certain the person in charge (let's called him Master Chef) was giving me a challenge. All fine and good, but I did not want to change my family stew recipe for the sake of a pork chop.

I ended up buying my own pork tenderloins and didn't touch the pork chops. Yes, Master Chef jeered at me when he saw me with my bags of tenderloins, saying "A chef has to be flexible. Pork chops can do a lot of beautiful dishes. Don't you know?" I simply told him: "Sorry Chef, but I am doing a pork stew dish tonight; it requires 45 minutes to cook; and that is the dish my clients are expecting." It's my show and it will be run my way.

The class went very well—as usual—and none of the clients knew about my pork chop saga. But that was the last time I taught at that school. I walked out, without turning back. No regrets!

# 葡萄牙白酒炆豬柳
## Portuguese Pork Tenderloin

我和朋友一起吃飯時，我喜歡為他們準備這道菜。這菜式的做法容易，也可以提前準備，吃的時候把它翻熱就可以了。我喜歡用一碟白飯來拌這豬柳，或者配北非小米Couscous也不錯，至少聽起來蠻健康……哈哈哈！

This is a dish that I love to prepare when having friends over for dinner. It is very easy to make and can be prepared in advance and just heated up when ready to serve. I love having a plate of white rice to complement the tenderloin, or couscous is not such a bad idea. It sounds healthy at least... Hahaha.

| 材料（4人份量） | | Ingredient (Makes 4 servings) | |
| --- | --- | --- | --- |
| 中筋麵粉 | 3湯匙 | All-purpose flour | 3 tablespoons |
| 鹽 | 2湯匙 | Salt | 2 tablespoons |
| 黑胡椒粉 | 1湯匙 | Black pepper | 1 tablespoon |
| 甜紅椒粉 | 1茶匙 | Paprika | 1 teaspoon |
| 豬柳 | 450克 | Pork tenderloin, trimmed | 450g |
| 無鹽牛油 | 2湯匙 | Unsalted butter | 2 tablespoons |
| 洋蔥切片 | 1個 | White onion, sliced | 1 |
| 啡蘑菇切片 | 250克 | mushrooms sliced | 250g |
| 白酒 | 160毫升 | Dry white wine | 160ml |
| 雞湯 | 150毫升 | Chicken broth | 150ml |
| 新鮮迷迭香 | ¼茶匙 | Fresh rosemary | ¼ teaspoon |
| 檸檬汁 | 1湯匙 | Lemon juice | 1 tablespoon |
| 新鮮蕃茜切碎 | 2湯匙 | Fresh parsley, chopped | 2 tablespoons |

做法

將麵粉、鹽、黑胡椒粉和辣椒粉放在碟上拌勻。將豬柳撲上已拌勻的麵粉。

在深煎鍋中以中火融化牛油。落豬柳，每邊煎約5分鐘，直至金黃色。加入洋蔥和蘑菇，炒1至2分鐘，直至聞到蘑菇的香味。

加入白酒、雞湯和迷迭香。蓋上蓋子，用細火煮30至40分鐘，直至豬柳變腍。在上菜前加入檸檬汁和蕃茜。

Combine the flour, salt, pepper, and paprika on a plate. Roll the tenderloin in the mixed flour until coated.

Melt the butter in a deep sauté pan over medium heat. Add the tenderloin and sear for 5 minutes on each side, until golden brown. Add the onion and mushrooms and sauté for 1 to 2 minutes, until you can smell the aroma of mushrooms.

Add the wine, broth, and rosemary. Cover the pan and cook over low heat for 30 to 40 minutes, until the pork is tender. Add the lemon juice and parsley just before serving.

葡萄牙白酒炆豬柳 *Portuguese Pork Tenderloin* 165

# 職業料理班學生

大家為什麼會參加我的葡國菜料理班？有的同學是因為對葡菜有濃厚的興趣，也有一部分是煮食迷，希望學到在香港較為少有的葡國菜和土生澳葡菜，也有一些同學是因為for the sake of enrolling。他們很多時候都忘記報了什麼課程，他們就是來上堂、拍照、打咭。當然有一些不會在家煮飯的一族，他們上課的時候順便把剛煮好的菜式拿回家「開飯」，間中也有一些煮食經驗是零的朋友。無論你是什麼緣故來參加我的料理班，我都是非常歡迎，來者不拒的。

大家有否聽過「職業料理班學生」這個term？原來香港的料理界中有一群職業學生，他們每天的工作就是出席不同的cooking class，然後把食譜售給其他媒體或cooking school。這批同學是受薪的。我自己也見過，我的嫲嫲葡國雞（p.73）在國內一些烹飪平台上出現，當然不是我提供食譜的。有時候這些同學也會在我的cooking class表現非常positive，不停的發問問題。我也親身體驗過同學們拿了我的食譜後在別的cooking school開班授課。

有一次有位同學由於知道我會開私人班（private class）課程，她便主動接觸我，向我查詢有關事宜。可能是我的警戒心放了下來，我立刻答應了她一個料理班的課程，並跟她合作，還給了她葡撻、葡式鷹咀豆炆牛肚和焗釀蟹蓋（p.121）的食譜做準備工作。那一個料理班在兩個月後才舉行，而那位同學大約在一星期後便跟我說project要告吹了。Oh damn！我給人騙了！！

後來我更知道原來那位同學開辦了一個類似的cooking class，旨在推介某一品牌的煮食用具。在當天，我的另一位同學也是座上客，後來我向那位同學後來索取了一份食譜看一看，那份食譜的錯別字跟我準備的一份exactly the same。我問了那位有出席料理班的同學如何，她跟我說「不行啊！」

經一事，長一智。哀哉！

# Professional Cooking Students

Do you know what is meant by "professional cooking students"? Well, among my clients, there are students who love cooking, especially Portuguese and Macanese home cooking; they are my fans. And there are people who enroll for the sake of enrolling, and often have no idea what I am teaching prior to the class. For them, they come to class, have a dish properly cooked, take nice photos, and the job is done. And there are also the newcomers, the ones who rarely cook at home. They want to experience Portuguese cooking and they are eager to learn. No matter what these students' agendas are, I try my best and deliver what I can to all my clients.

But then there are the professional cooking students, the ones who want my recipes. In class, some are silent and never ask questions; their mission is simply to collect recipes to sell to other channels. (Oh yes, I have seen my Avo Chicken (p.73) recipe in a mainland cooking platform.) The other type of professional student will ask me a lot of questions, learn my dishes, and then use them in their own cooking class—usually within just a week or month.

I had one student come to me at the end of class, and ask me about teaching in a business affiliate group in two months' time. I agreed and sent her my recipes for Portuguese Egg Tarts and Tripe in Porto Style, and Baked Crab Meat Shells (p.121). Less than a week later, she called me and said the event had to be cancelled because of "no registrations." I was like, "Wait... but the class is still two months away from now." Smells fishy, right?

Not surprisingly, not soon after a client told me about attending a Portuguese cooking class featuring a baked crab meat recipe and she said the class was really bad. The cooking instructor kept referring to the recipe during the class, and when asked about the origin of the dish, he was silent and could only bluff something stupid. I asked her to show me the recipe. Yes: It was my recipe (I know because there was a typo that was the same as in my recipe). Of course, the cooking class had been held at that same business affiliate venue.

What can I say? Sigh.

# 葡撻
## Portuguese Egg Tarts

大部分人都吃過葡撻。你知道為什麼我很少在課堂上教授做葡撻？首先，網上已經可以找到很多葡撻食譜，而一些更是有名的blogger（他們提供的食譜很好，儘管不是正宗葡萄牙式做法）去撰寫。傳統的葡撻做法實在太甜，我們要知道正確食葡撻的方式。

所以希望你跟著我建議的方式去做：正宗葡撻食譜會加入肉桂和檸檬皮，並且要趁熱食，至少趁它還是暖和的時候享用。千萬不要吃冷的或者是室溫的葡撻。配上一杯優質的黑咖啡（或濃咖啡），這是葡撻的最佳伴侶。也千萬不要把葡撻配上可口可樂或凍奶茶，這是對葡撻的侮辱。就像意大利人從不把菠蘿放在披薩上一樣道理。讓我們一起做個百分百的正宗葡撻吧！

哦！是的，這個食譜我用上急凍酥皮，因為急凍酥皮更容易在家裡處理。

Everyone knows Portuguese egg tarts. But do you know why I seldom teach the recipe in my classes? First, there are already a lot of recipes available on the web, by some very remarkable bloggers (their recipes are good, though not the proper Portuguese way). Second, the traditional recipe for egg tarts is way too sweet if you don't know how to make them properly.

So here is what I'd like you to do: Try my Portuguese egg tart recipe—it's authentic, with cinnamon and lemon zest—and be sure to enjoy your egg tarts while they are still hot, or at least warm. Never serve an egg tart that is cold or at room temperature. And a nice cup of black coffee (or espresso) is the best buddy to egg tarts. Don't even think of Coca-Cola or iced milk tea. It's an insult to Portuguese egg tarts. It's similar to how Italians don't put pineapple on pizza. Let's do it the Portuguese way.

Oh yes, I am using frozen puff pastry, as it's far easier for everyone to handle at home.

材料（10個葡撻份量）

| | |
|---|---|
| 白糖粉 | 180克 |
| 水 | 300毫升 |
| 烹調用忌廉 | 250毫升 |
| 肉桂條 | 1條 |
| 檸檬皮 | ½個檸檬 |
| 雲呢拿油 | ½茶匙 |
| 蛋黃 | 6隻 |
| 牛奶 | 50毫升 |
| 無鹽牛油，雪凍和切粒 | 掃盤用 |
| 急凍酥皮，解凍 | 2張 |
| 肉桂粉 | 2茶匙 |

Ingredient (Makes 10 tarts )

| | |
|---|---|
| Caster sugar | 180g |
| Water | 300ml |
| Cooking cream | 250ml |
| Cinnamon stick | 1 |
| Peeled lemon zest | from ½ lemon |
| Vanilla extract | ½ teaspoon |
| Egg yolks | 6 |
| Milk | 50ml |
| Unsalted butter, frozen and cubed | for greasing |
| Frozen puff pastry, thawed | 2 sheets |
| Ground cinnamon | 2 teaspoons |

做法

將白糖粉和水放入平底鍋中攪勻，用中火煮滾。然後繼續用細火加熱約5分鐘，直至糖完全溶解。將糖漿放在一邊放涼。

在第二個平底鍋中加入忌廉、肉桂條、檸檬皮和雲呢拿油，攪勻，用中火煮5分鐘，直至你能聞到檸檬和肉桂的清香。但不要讓它煮至大滾。

將糖漿加入忌廉中，用中火煮至剛好沸騰。將蛋黃與牛奶拌勻，倒入剛煮滾的忌廉裡。用中火邊煮邊攪拌忌廉，直至稍微杰身，約2至3分鐘。弄掉檸檬皮和肉桂條。離火，讓吉士醬冷卻。篩走氣泡、蛋塊和未融化的忌廉粒 。這時候吉士醬應該看起來美味滑溜。放在一邊待涼。

焗爐在200°C預熱15分鐘。用牛油塗抹10個蛋撻模。

How to cook

Combine the caster sugar and water in a saucepan and stir well. Bring to a boil over medium heat. Let it continue to boil on low heat for about 5 minutes, until the sugar has dissolved completely. Set the sugar syrup aside and let cool.

Combine the cream, cinnamon stick, lemon zest, and vanilla in a second saucepan. Whisk well over medium heat for 5 minutes, until you can smell a light aroma of lemon and cinnamon. Don't let it burn.

Add the sugar syrup to the cream and cook over medium heat until it starts boiling. Mix the egg yolks with the milk, and stir into the cream mixture. Stir over medium heat until it thickens slightly, 2 to 3 minutes. Remove the lemon zest and cinnamon stick. Remove the pan from the heat and let the custard cool. Strain through a sieve to remove any bubbles and rough egg or

酥皮通常是一張張的。取出2張解凍的酥皮，然後將每張酥皮緊緊地捲成卷狀。將每個卷切成5份，3至4厘米寬，並搓成條狀。

將每個酥皮放入模中，用手指輕輕地在酥皮上向下壓，不要破壞酥皮皮層。你的手指可能需要塗上一些牛油，好讓你更容易將酥皮壓到2至3毫米厚，並鋪貼整個撻模。

將吉士醬僅僅填滿整個撻模，不要過滿。焗20至25分鐘，直至凝固。此時你要留意在焗爐內的蛋撻：吉士醬升起，並在表面形成一層深啡色的薄膜。別打開焗爐。

讓蛋撻留在焗爐冷卻5分鐘。在蛋撻仍然熱哄哄的時候灑上肉桂粉。

謹記葡撻要在熱騰騰或暖呼呼的時候享用！

cream bits. The custard should look deliciously silky by now. Set aside.

Preheat the oven at 200°C for 15 minutes. Grease 10 cups of a standard muffin pan with the frozen butter.

The pastry usually comes in sheets. Take 2 thawed sheets and roll each up into a tight roll. Cut each into 5 pieces, 3 to 4cm-wide, and separate into strips.

Place each piece of pastry in a muffin cup. Using your fingers, gently press the pastry downwards into the pan without breaking the layers. This is important as these layers give us those hundreds of layers of flaky pastry—so let's not break them. You might want to rub your fingers with some butter and press the pastry to fit the muffin cups so it is about 2 to 3mm thick.

Fill each pastry cup with the egg custard until it's almost filled, but not fully. Bake for 20 to 25 minutes, until set. Watch the tarts: They should rise and form a very dark brown skin on the surface. That's expected. Leave them alone. Do not open the oven door during this time.

Let the tarts cool for 5 minutes in the oven. Sprinkle with cinnamon while they are still hot. Start digging into one while they are fresh and warm! Enjoy!

1. 盡量不要打開焗爐檢查蛋撻。不斷開關焗爐會讓蛋撻蛋醬上升和下塌，導致撻面出現裂縫。

2. 翻熱蛋撻，將焗爐預熱至160°C，然後加熱蛋撻10至15分鐘。

1. Try not to open the oven to check on the tarts. Constant opening and closing of the oven door can cause the tarts to rise and sink, resulting in cracks on the top.

2. To warm up cold tarts, preheat the oven to 160°C and heat the tarts for 10 to 15 minutes.

# 鷹咀豆煮牛肚
## Tripe, Porto Style

說到牛肚，大家就會想到經典港式牛腩湯河粉，或者小販檔賣的牛肚，配中國芥茉和辣椒醬。你可知道葡萄牙人也常吃牛肚嗎？這是我的家傳食譜波爾多牛肚。波爾多是葡萄牙北部的一個大型海港，那裡的人會用這個菜來拌白飯，配上簡單的沙律、麵包和一杯紅酒。

When it comes to tripe, one will think of the typical flat rice noodle soup with Hong Kong style tripe（河粉）, in soup, or the hawker stalls of braised tripe served with Chinese mustard and chili sauce. But did you know that tripe is also commonly eaten in Portugal? This is my family version of Porto style tripe. Porto is a large seaport in north Portugal, and people there serve their tripe with steamed rice, a simple salad, bread, and a glass of red wine.

| 材料（4人份量） | | Ingredient (Makes 4 servings) | |
|---|---|---|---|
| 牛肚 | 500克 | Tripe | 500g |
| 洋蔥切成4份 | 1個 | White onion, quartered | 1 |
| 雞湯 | 940毫升 | Chicken broth | 940ml |
| 橄欖油 | 4湯匙 | Olive oil | 4 tablespoons |
| 洋蔥切碎 | 1個 | White onion, minced | 1 |
| 紅蘿蔔去皮切粒 | 1隻 | Carrot, peeled and diced | 1 |
| 蒜頭切片 | 6瓣 | Garlic cloves, sliced | 6 |
| 蕃茄切碎 | 1至2個 | Tomatoes, chopped | 1 to 2 |
| 西班牙腸或葡萄牙腸切成小塊 | 120g | Chorizo or | |
| 砵酒 | 45毫升 | Portuguese sausage, diced | 120g |
| 月桂葉 | 3塊 | Tawny port wine | 45ml |
| 罐裝鷹咀豆 | 128克 | Bay leaves | 3 |
| 白酒或紅酒醋 | 2湯匙 | Chickpeas (canned) | 128g |
| | | White or red wine vinegar | 2 tablespoons |

做法

將牛肚和洋蔥（切成4份）放在一個大煎鍋裡，放入雞湯，蓋上蓋子。用中高火將湯煮滾。將火調細，煮30至45分鐘，直至牛肚變臉。冷卻後，將牛肚隔汁並切成3厘米的方塊。保留煮牛肚的雞湯。

用中火加熱鑄鐵鍋或不黏砂鍋中的橄欖油。加入切碎的洋蔥，炒約5分鐘，直至軟身。加入紅蘿蔔、蒜頭、蕃茄和肉腸，再煮10分鐘，直到紅蘿蔔變軟，蕃茄煮爛。

倒入預留的雞湯、砵酒和月桂葉。加入牛肚、鷹咀豆和醋，用細火炆90分鐘，直到牛肚完全吸收汁液，變得入味。

這道菜式不會有太多醬汁，但如果在烹煮過程中變得乾涸，可以按個別情況下加添雞湯。

謹記吃時要弄走月桂葉。趁熱享用！

How to cook

Place the tripe and onion quarters in a large saucepan and cover with the broth. Bring to a boil over medium-high heat. Reduce the heat and cook until the tripe is tender. This normally takes 30 to 45 minutes. When cool, drain and cut the tripe into 3cm squares. Reserve the chicken broth.

Heat the olive oil in a cast iron Dutch oven or nonstick casserole over medium heat. Add the minced onions and sauté for about 5 minutes, until soft. Add the carrot, garlic, tomatoes, and chorizo and cook for another 10 minutes, until the carrot turns softer and the tomatoes are breaking apart.

Pour in the reserved chicken broth, the port, and bay leaves. Add the tripe, chickpeas, and vinegar and simmer over low heat for 90 minutes, until the tripe is very tasty after absorbing the liquid.

There should not be a great deal of liquid left, but add more broth or water if needed during the cooking to prevent burning.

Remove and discard the bay leaves and serve hot. Enjoy!

鷹咀豆煮牛肚 *Tripe, Porto Style*  177

# 最長的一課

來上過我料理班的同學都會主動跟我說氣氛良好，大家上堂有講有笑的。的而且確，我一直都是希望同學們可以開開心心來上堂，在做菜的時候可以暫時放下家裡的問題，又或是工作上的煩惱。所以我每一次都是笑面迎人，希望大家也被我的positive energy感染到。

不過我始終都是一個有血有肉的人。還記得有一次伴我超過17年的毛孩走了，而我立即處理牠的身後事便要趕回cooking school上堂，那一天我的心情跌到最低點，眼看著3隻年長的毛孩在8個月內一個一個的離開了，EQ怎樣高也難掩傷痛之情。

而我也不打算告訴我的同學們，只是靜悄悄告訴cooking school便是了。

上課的時間到了，一切還ok啊！但突然有一位同學從手袋拿出她的手機，然後跟大家分享了她的毛孩在游泳的video。她當然不知道發生了什麼事，而我的眼淚已經hold不住的流了出來。我就唯有默默的切洋蔥，同學們就開始把話題由煮牛膝轉為夏天照顧愛犬之談。同學們都知道我是一名愛狗之人，所以她們也非常熱烈和高興的問我有關狗狗飲食的問題。天啊！我的腦海裏就是浮現著在不夠24小時前的景象：我的手提著毛孩的手，然後我在牠耳邊說「Thank you baby for watching and taking care of me in these 17 years. You have done enough. Walk towards the light.」。然後Doctor Harrison就為毛孩哥哥注入了最後的一針，而牠的呼吸聲也由抽畜轉為細聲，最後靜止了。再聽不到毛孩的呼吸聲，只有我的哭泣聲，完了⋯⋯

那一堂是最長的一課！

# Toughest Time

I believe that, for many of my clients, there's a hidden agenda behind taking a cooking class: therapy and relaxation. I never ask my clients why they sign up for a class. I only want to deliver the best I can, and make sure they enjoy the class and take home not just delicious food, but a positive and happy vibe.

So it's important to create that positive vibe throughout the entire class. I leave all my critical home affairs outside the kitchen during class. I only talk about positive things, and rarely anything negative. Do I sound like I'm coming from a reiki workshop or ashtanga yoga? LOL.

But there was one day in particular when I was really down and sad: My beloved 17-year-old doggie had passed away the day before. (In fact, I lost three of my beloved dogs in just eight months between 2017 and 2018.) No one knew about it, only the cooking school staff.

During the class, one of my clients took out her phone to show a video of her dog swimming happily in a pool. Everyone was smiling, laughing, and talking. She didn't mean anything by it, of course. My colleagues from the school looked

at me, and saw the tears rolling down my cheek. Without letting anyone know, I continued with my chopping and cooking, and kept talking to another client. So that was the first time I cried because of chopping onions. But the even harder part is that during the rest of the class the talk was of keeping dogs, diets for dogs, everything dogs. I had to participate in the conversation because everyone knew I am a dog lover, but they didn't know what I'd been through in the last 24 hours:

The moment when I was holding the paw of my baby in my hand, whispering into his ears and saying, "Thank you baby for watching and taking care of me in these 17 years. You have done enough. Walk towards the light." And as I was murmuring these words, I saw Dr. Harrison injecting the last fluid into the drip, and baby's breathing became weaker and lighter, and he was gone. No more heavy breathing from Baby.....silence....but for my crying.

That was the longest cooking class I had, and it was the toughest.

# 紅酒炆牛仔膝

Braised Veal Shanks with Warm Spices

這個食譜改良自經典的葡國香草牛腱。朋友到我家裡吃飯時，我會做這個菜餚給他們品嚐，我喜歡用牛仔膝代替牛仔肉來煮這個菜，因為牛仔膝更嫩更具口感。你可改用牛腱或牛尾來做一頓豐富的家常菜，而牛仔膝則用來宴客。這道菜拌白飯或薯蓉就是最佳的配搭.

This recipe is adapted from a classic Azorean braise of beef shins and spices. I like to use veal shanks instead of veal when having friends over for dinner, because veal is more delicate, tender, and exquisite. But feel free to use beef shin or oxtail for a hearty family meal. In any case, serve with white rice or mashed potatoes—that's how I love the shanks.

| 材料（4人份量） | | Ingredient (Makes 4 servings) | |
|---|---|---|---|
| 黑椒粒 | 1湯匙 | Whole black peppercorns | 1 tablespoon |
| 丁香 | 6粒 | Whole cloves | 6 |
| 肉桂條 | 1根 | Cinnamon stick | 1 |
| 八角 | 3粒 | Whole star anise | 3 |
| 月桂葉 | 2片 | Bay leaves | 2 |
| 厚切煙肉（每粒1厘米） | 170克 | Slab bacon, cut into 1cm cubes | 170g |
| 橄欖油 | 2湯匙 | Olive oil | 2 tablespoons |
| 牛仔膝（每塊3厘米厚） | 4塊 | Veal shanks, 3cm-thick cut | 4 |
| 鹽和胡椒粉 | 適量 | Salt and pepper | to taste |
| 洋蔥切粒 | 2個 | White onions, diced | 2 |
| 蒜頭切碎 | 6瓣 | Garlic cloves, minced | 6 |
| 紅酒 | 500毫升 | Red wine | 500ml |
| 牛肉湯 | 600毫升 | Beef broth | 600ml |
| 黑糖蜜*（見下廚小貼士） | 1湯匙 | Molasses* (see Tip) | 1 tablespoon |

做法

將黑椒粒、丁香、肉桂條、八角、月桂葉放入芝士布，然後包紮起或放在魚湯袋中, 準備待用。

用中低火加熱平底鍋或鑄鐵鍋。加入煙肉並煎至出油及酥脆，大約10分鐘。然後放在餐巾紙上吸走多餘油份並放在一邊待用。

將橄欖油加入煎鍋，中高火加熱。用鹽和胡椒粉醃牛仔膝，然後放在煎鍋中煎至變金黃色，每邊約5分鐘。轉移到碟上並放在一邊。

煎鍋火候調低至中火。將洋蔥炒至軟身，約7分鐘。加入蒜頭，再煮1分鐘。加入紅酒、牛肉湯、黑糖蜜和1茶匙鹽，煮滾。將煙肉和牛仔膝及所有剩餘肉汁一併倒回鍋中。放入香料包。蓋上蓋子並用細火炆一個半小時左右，直至可將叉插入牛仔膝。

離火。用漏勺將肉轉到碟上，如肉和骨已煮至分開，這效果最為理想。

將煎鍋內的香料包弄掉。如有必要，用大火將醬汁減少至250毫升或更少。

撇清醬汁中的油脂，用鹽和胡椒粉調味。

將肉和骨頭（以及骨髓）分成四份上碟，然後淋上醬汁。

3-2-1，可以享用了！

How to cook

Tie the peppercorns, cloves, cinnamon stick, star anise, and bay leaves into a cheesecloth pouch or fish broth pouch. Set aside.

Heat a large saucepan or Dutch oven over medium-low heat until hot. Add the bacon and sizzle to render the fat and crisp up the meaty bits, about 10 minutes. Transfer to paper towels to drain and set aside.

Add the olive oil to the fat in the pan and heat over medium-high heat. Season the shanks well with salt and pepper, then sear in the pan until well browned, about 5 minutes per side. Transfer to a plate and set aside.

Lower the heat to medium. Add the onions to the same pan and sauté until softened, about 7 minutes. Add the garlic and cook for 1 minute more. Add the wine, broth, molasses, and 1 teaspoon salt and bring to a boil. Return the bacon and the shanks, along with any accumulated juice, to the saucepan. Nestle in the spice pouch. Cover and braise over low heat for about 1½ hours, until the meat is fork-tender.

Remove the saucepan from the heat. Using a slotted spoon, transfer the meat to a plate; it should be falling apart, that's what you are aiming for. Transfer the bones too.

Discard any rubbery bits still clinging to the meat and toss out the spice pouch. If necessary, reduce the cooking liquid over high heat to about 250ml or less. Skim the fat from the sauce and season with salt and pepper to taste.

Divide the meat and bones (and their marrow) among four plates and ladle the sauce on top.

3-2-1, enjoy!

 你可以用20克冰糖代替黑糖蜜。
You can substitute 20g of rock sugar for the molasses.

# 木糠布甸
## Serradura

Serradura葡文的意思解作為木糠，而木糠布甸也被稱為澳門布甸。這個名字是形容布甸上餅乾碎的樣子。木糠布甸 的做法製非常容易而且超級美味。這是一個原創食譜，它之所以是原創是因為這是我的家傳食譜，也是嫲嫲傳授給我的。我喜歡隨意在布甸最上層放oreo餅碎，也可以放綠茶粉。在製作甜品之前，記得要把用來打忌廉的不銹鋼碗放在雪櫃至少1小時。

Serradura, the Portuguese word for "sawdust," is also known as Macau pudding and sawdust pudding. The name refers to the way the biscuit crumbs look in the pudding. It's so easy to make and super yummy. But watch out for your waistline as this is truly a devil dessert. I call it the original. It's my family recipe and how Avo taught me. Feel free to top the pudding with Oreo crumbs, or mix in green tea powder if you like. Be sure to put the bowl for whipping the cream in the freezer at least an hour before making the dessert.

材料（4人份量）

| 瑪麗餅 | 200克 |
|---|---|
| 淡忌廉（製作甜品用） | 200毫升 |
| 白糖粉 | 2湯匙 |
| 煉奶 | 1湯匙 |
| 雲呢拿油 | 1茶匙 |

Ingredient (Makes 4 servings)

| Marie biscuits | 200g |
|---|---|
| Whipping cream (for dessert) | 200ml |
| Caster sugar | 2 tablespoons |
| Condensed milk | 1 tablespoon |
| Vanilla extract | 1 teaspoon |

做法

用攪拌機將瑪麗餅磨成幼細的餅乾碎。放在一邊待用。

從雪櫃取出不銹鋼碗,用手持式攪拌器以高速攪拌忌廉。當忌廉打至挺身,便加入白糖粉、煉奶和雲呢拿精油,繼續攪拌,直到全部忌廉軟滑。

準備4隻你心儀的布甸杯,放一層忌廉在杯內。加上一層瑪麗餅乾碎。重複加上一層忌廉和餅碎,要做的層數可以按個人喜好及布甸杯的大小而定,但最上一層會放瑪麗餅碎或oreo餅碎。

完成後,將布甸杯蓋上保鮮膜放入雪櫃冷卻1小時。

1個小時後,請慢用!

How to cook

With a food processor, grind the Marie biscuits into fine crumbs. Set aside.

In a very cold stainless-steel bowl, use a handheld mixer to whip the cream at high speed. Once the cream has peaks, add the caster sugar, condensed milk, and vanilla and continue whipping until the mixture reaches a soft, smooth, and even consistency.

Pour a layer of the cream into each of four serving glasses of your choice. Add a layer of the Marie biscuit crumbs. Repeat the layers of cream and crumbs, making as many layers as you like, but end with a layer of crumbs. (Or you may like to top with Oreo crumbs.)

Once done, covered with cling wrap and place the glasses in the fridge for 1 hour to chill.

An hour later, enjoy!

Tips 怎樣處理剩下的瑪麗餅餅碎?你可以把餅碎拌入雲呢拿雪糕或希臘式乳酪。
這是另一種邪惡式甜品!
What can you do with any remaining unused Marie biscuits crumbs? Mix them with vanilla ice cream or Greek yogurt. Just another devilish indulgence!

# 矛盾

在最初的時候，我就在一些飲食平台上post一些食譜，如Cook1Cook，然後在Wordpress開了一個blog，還記得首個分享食譜便是Almôndegas葡式肉丸，後來我也展開了開辦料理班的課程，那麼我繼續post食譜還是堅決做cooking class的原則呢？Well，矛盾開始出現了。

看見很多前輩，每天努力的去出post，分享食譜，天天如是。對啊！他們的粉絲人數都在5位數字，找他們做news feed的sponsor多得滿瀉。但我再看看他們的engagement rate，是10%也不足，而且每天就是post食譜，跟粉絲們沒有甚麼互動的啊！

那我要粉絲人數，還是要有質素的內容嗎？我最後還是選擇了後者，而最重要的我仍然相信two-way communication的重要性，一場料理班才是我喜歡分享煮食經驗的channel，但是我如何令到更多朋友知道我的「存在」呢？免費食譜？收費料理班？這兩個互相抗衡的問題all along都出現在我的腦海裏。

從小到大我都非常熱愛閱讀，也喜歡寫一些creative writing，我便開始嘗試寫一些有關菜式背後的故事，一些生活的點滴，介紹餐廳的文章，後來還報了一些課程學拍片，剪片。哈哈，粉絲們的反應也不錯啊！就這樣，我開始做了一位slashie（即烹飪導師／blogger／西餐達人／生活家／作家），有媒體朋友更稱我為influencer，而不是KOL。

要做一位稱職的料理導師is never easy，要做一位成功的blogger或是KOL／influencer就是更難。而我就是一位多重身份的slashie，真的難上加難，今天我的書已經上架了，那麼下一步是甚麼呢？這個social media真的天天在變，當我開始掌握如何善用我的Facebook page，突然又走出一些IGTV的東西。真的是Oh My God！

# *Dilemma*

In 2016, I began to find my postings on social media to be a bit meaningless: just travel photos of places that I have been around the world. I wanted to do something different, perhaps something more meaningful. So I started posting cooking photos on social media, sharing my recipes and stories. The very first recipe I posted on Wordpress was Almôndegas—Portuguese meatballs. Then I started teaching cooking classes, and created my Facebook page, and migrated my blog from Wordpress to Blogspot. In a way, I have become both a cooking instructor and a blogger. And here's my dilemma: Do I concentrate on cooking classes or on social media?

There are a lot of well-known cooking bloggers in town, and they are so hard-working and dedicated, posting recipe after recipe on social media every day. And the recipes are all free. Yet, these bloggers aren't cooking instructors. I can do the same (or, okay, I can post twice a week, to be honest). But the more I post on my blog or Facebook, the more difficult it is for people to agree to pay $600+ to take one of my classes. I prefer the format of cooking class as I love the interactions with people. I enjoy getting to know my clients, chatting with them and exchanging cooking experiences. On social media, just seeing the "Like"s and emoticons pile up just isn't enough for me (I know I am greedy). I want more.

No one can deny the power of social media. My Facebook page is a tool to reach more people, and let them know about my cooking classes and home recipes. Isn't this what I want? How do I get people to know who I am? More free recipes seem to be the solution. I get a lot of private messages every week asking me for my recipes. And sometimes it's so difficult to say no with a smile.

My challenge is how to balance providing free recipes and my paid cooking classes. But mostly, I don't think about it too much and simply just go with the flow. I've also started writing reviews of restaurants that I visit, and short journals about my thoughts, and people and fans love them. So I gradually cut down posting recipes on other social media platforms. All seems to be working well because the number of followers and LIKES are on the increase every week.

It's never easy to be a good instructor and a blogger. I am working so hard. But friends have already told me I am not just a blogger now, but a KOL (key opinion leader) and Influencer. Which means that I need to know terms like organic reach, and engagement rate, and trends of IGTV. It's really OMG! So here's the first recipe I blogged.

# 迷你漢堡扒
## Almôndegas

我發覺葡萄牙料理跟意大利料理在烹調方式上非常相似。例如，葡萄牙和意大利菜都會用上大量的蕃茄、洋蔥和蒜頭。Almôndegas葡萄牙語是肉丸的意思，通常會將肉丸配上蕃茄醬，猶如意大利菜的meatballs pasta。我換個方式，把它變成了迷你漢堡扒。作為一個頭盤或派對菜式是不錯的選擇。

The more that I look into the history of Portuguese cooking, the more I find it very similar to Italian cooking. For example, both Portuguese and Italian cooking use a lot of tomato, onions, and garlic. And almôndegas, which in Portuguese means meatballs, is usually presented like meatballs with tomato sauce. I changed it and turned it into mini burgers. It's a good party food option.

| 材料（12個迷你漢堡扒份量） | | Ingredient (Makes 12 mini burger steaks) | |
| --- | --- | --- | --- |
| 免治豬肉 | 250克 | Minced pork | 250g |
| 高達芝士磨碎 | 50克 | Gouda cheese, grated | 50g |
| 青橄欖切碎 | 5粒 | Green olives, finely chopped | 5 |
| 鹽和胡椒粉 | 適量 | Salt and pepper | to taste |
| 雞蛋發勻 | 1隻 | Egg, beaten | 1 |
| 日本麵包糠 | 2湯匙 | Japanese panko | 2 tablespoons |
| 橄欖油或米糠油 | 用於淺煎炸 | Olive oil or rice bran oil | for shallow frying |

做法

將豬肉、芝士和橄欖在不銹鋼碗中攪拌並搓勻至厚實滑身。用鹽和胡椒粉調味。

製成12個小肉球。將每個球浸入已打好的蛋漿，然後撲上麵包糠。將肉球弄扁平成一個直徑約5厘米的小肉餅。

在不黏鍋的煎鍋中放4厘米油，用中火加熱。分批煎肉餅，反轉一次，煎約5分鐘，直到煎熟，外層帶點微焦和酥脆。

熱食或放涼後食均可。

How to cook

Combine the pork, cheese, and olives in a stainless bowl. Knead together until the mixture is smooth and thick. Season with salt and pepper.

Form the mixture into 12 small balls. Dip each ball in the beaten egg, then dredge in the panko. Flatten each to form a small patty about 5cm in diameter.

Heat about 4cm oil in a nonstick frying pan over medium heat. In batches, add the patties and cook, turning once, for about 5 minutes, until cooked through and a bit brown and crispy on the outside.

Serve hot or at room temperature.

 你可以全部用上免治牛肉，或者將100克免治牛肉和150克免治豬肉混合在一起。

You can use minced beef if you like, or simply combine 100g minced beef and 150g minced pork.

# MENU

# A餐？B餐？

從小到大，我都很喜歡吃港式西餐，尤其是喜歡看見不同菜式被劃分為A餐、B餐、C餐……每次去到港式西餐廳，我總是喜歡向侍應點紅湯或白湯。說真的，香港式的紅湯和白湯真的幾好飲（不要笑我）！

寫到這裡，我突然想做一個以葡萄牙菜為主的A餐和土生澳葡菜為主的B餐。

|  A餐  |  B餐  |
| :---: | :---: |
| 砵酒吞拿魚醬 | 蝦多士 |
| 烤蕃茄湯 | 豬肘骨 |
| 煎沙甸魚配葡式辣醬 | 蠔油青檸煮雞 |
| 葡萄牙蒜蓉雞 | 西洋一品鍋（p.113） |
| 葡萄牙法式多士 | 葡式甜飯 |

要去得盡興一點，大家可以在餐後飲一杯black coffee（葡萄牙人稱為Bica）又或者一小杯砵酒，然後聽住Amalia Rodrigues的Fado音樂。

樂也！

# Suggested Menus

So how about we have a Portuguese themed set menu or a cozy home-cooking Macanese dinner? I have listed two different sets of menus and hope they will be a good start for "something Portuguese."

### Set A

Tuna Spread

Roasted Tomato Soup

Grilled Sardines over Portuguese Stew

Portuguese Garlic Chicken

Rabanadas
(Portuguese French Toast with Port)

### Set B

Shrimp Toast

Cotovelo de Porco (Pig Elbow Bone)

Chicken with Oyster Sauce and Lime

Daibo (p.113)

Creamy Portuguese Rice Pudding (Arroz Doce)

And to finish off with the meal, it's best to have a cup of black coffee (Bica is the name they use in Portugal) or just a small glass of port wine. Music ? Hmmmm, Fado of course. Amalia Rodrigues is my favorite Fado singer.

# 砵酒吞拿魚醬
## Tuna Spread

當你去葡萄牙的餐館用膳時，通常侍應生會先給你端上一些前菜，例如Azeitonas mistas（橄欖拼盤），Pasta de atum（吞拿魚醬）和Papo-secos（乾麵包）。既經典又普遍的吞拿魚醬本應塗在麵包、多士的一角或餅乾上，但我則喜歡吃得瀟灑一點：用匙羹點來吃！

When you go to a restaurant in Portugal, you're usually first served a few small dishes of azeitonas mistas (mixed olives), pasta de atum (tuna spread), and papo-secos (dried bread). The classic all-time-favorite tuna spread should be smeared on bread, toast points, plain crackers, or simply dipped into with a spoon—like I do!

| 材料（2杯份量） | | Ingredient (Makes 2 cups) | |
|---|---|---|---|
| 橄欖油 | 2湯匙 | Olive oil | 2 tablespoons |
| 洋蔥切碎 | ½個 | White onion, minced | ½ |
| 蒜頭切碎 | 2瓣 | Garlic cloves, minced | 2 |
| 油浸吞拿魚 | 2罐（120克） | Tuna in olive oil (canned) | 2 cans (120g) |
| 無鹽牛油 | 100克 | Unsalted butter | 100g |
| 砵酒 | 3湯匙 | Tawny port wine | 3 tablespoons |
| 新鮮檸檬汁 | 1湯匙 | Fresh lemon juice | 1 tablespoon |
| 鹽和胡椒粉 | 適量 | Salt and pepper | to taste |

做法

於小煎鍋裡，用中火加熱橄欖油，直到微滾。放入洋蔥，經常攪拌直至洋蔥光滑並呈深金黃色，約5分鐘。加入蒜頭，煮1分鐘。然後離火於室溫冷卻。

將預先製作好的洋蔥和蒜頭與吞拿魚、牛油、砵酒和檸檬汁倒入攪拌機拌勻，直至形成滑身的醬。用鹽和胡椒粉調味。

## How to cook

Heat the olive oil in a small skillet over medium heat until it shimmers. Drop in the onion and cook, stirring often, until slick and deeply golden browned, about 5 minutes. Add the garlic and cook for 1 minute. Remove from the heat and let cool to room temperature.

Combine the onion mixture with the tuna, butter, port, and lemon juice in a food processor and buzz until a smooth paste is formed. Season with salt and pepper.

Dig in and enjoy!

將吞拿魚醬緊緊封蓋好，放入雪櫃，可保存一星期。但我懷疑有幾多人能保留一個星期？因為這個醬確實太誘人，大家會忍不住天天吃，很快把它吃光。哈哈！

The spread will last for 1 week, tightly covered, in the fridge. But I doubt if you can resist the temptation to dip into it daily.

# 烤蕃茄湯
## Roasted Tomato Soup

因為我的烹飪課程一般長2個半到3個小時，所以有些菜餚很難在課堂上分享。烤蕃茄湯是一個例子：它的做法並不難，但需要一些時間去烤蕃茄。有空的話不妨試一試，我相信你的朋友和家人會更加喜歡你！

Because my cooking classes are usually 2½ to 3 hours long, there are some dishes that are difficult to share in the class format. Roasted tomato soup is one: It's not difficult at all, but it takes some time, especially for roasting the tomatoes. Try this out when you have time, and I am sure friends and family will love you more and more!

| 材料（4人份量） | | Ingredient (Makes 4 servings) | |
| --- | --- | --- | --- |
| 新鮮蕃茄 | 500克 | Fresh vine tomatoes | 500g |
| 橄欖油 | 2湯匙 | Olive oil | 2 tablespoons |
| 糖 | 1茶匙 | Sugar | 1 teaspoon |
| 紅辣椒碎 | ½茶匙 | Crushed red pepper | ½ teaspoon |
| 洋蔥切碎 | 2個 | White onions, chopped | 2 |
| 蒜頭去皮和壓扁 | 2瓣 | Garlic cloves, peeled and crushed | 2 |
| 甜紅椒粉 | 2茶匙 | Paprika | 2 teaspoons |
| 罐裝蕃茄 | 2罐（430克） | Plum tomatoes (canned) | 2 cans (430g) |
| 雞湯 | 1.5升 | Chicken broth | 1.5 liters |
| 蝦 | 4隻 | Shrimps | 4 |
| 鹽和胡椒粉 | 適量 | Salt and pepper | to taste |

## 做法

將焗爐預熱至100°C。在大烤盤上噴上不黏模噴霧或掃上橄欖油。

將新鮮蕃茄縱向切成兩半。在烤盤上單層排列，切面朝下。淋上1湯匙橄欖油和糖。焗2個小時，直到非常軟身和起皺。放在一邊放涼。

在燉鍋或湯鍋中，用中火加熱剩餘的1湯匙橄欖油和紅辣椒碎。加入洋蔥和蒜頭，邊攪拌邊煮，直至淡微微金黃色，約10至12分鐘。加入甜紅椒粉，煮1分鐘。加入罐裝蕃茄和雞湯，並用湯匙壓碎蕃茄。煮沸後調至細火，半開鍋蓋，繼續煮20分鐘，直到蕃茄變為稀欄。離火並放涼。

同時，準備滾水，落鹽，放蝦煮1分鐘，直到煮熟。將蝦瀝乾並待涼，然後剝殼。

將湯倒入碗中，隔出湯渣並以攪拌機打碎，直至滑溜。必要時加入少許湯或水。將攪拌後的湯蓉和湯放回鍋中拌勻。

用攪拌機將已焗好的蕃茄攪成蓉，然後過粗篩（弄掉隔出的蕃茄）。最後將蕃茄蓉和湯拌勻，用鹽和胡椒粉調味。

將湯放入碗中，放上1隻蝦作裝飾。即可享用。

## How to cook

Preheat the oven to 100°C. Oil a large baking sheet with nonstick spray or olive oil.

Cut the fresh tomatoes in half lengthwise. Arrange, cut-side down, in a single layer on the baking sheet. Drizzle with 1 tablespoon of the olive oil and the sugar. Roast for 2 hours, until very soft and wrinkled. Set aside and let cool.

In a Dutch oven or soup pot, heat the remaining 1 tablespoon olive oil and crushed red pepper over medium heat. Add the onions and garlic and cook, stirring, until lightly browned, 10 to 12 minutes. Add the paprika and cook for 1 minute. Add the canned tomatoes and broth and use a spoon to break up the tomatoes. Bring to a boil, reduce the heat to low, and simmer, partially covered, for 20 minutes, until the tomatoes are falling apart and smeared. Remove from the heat and let cool slightly.

Meanwhile, cook the shrimp in a large saucepan of boiling salted water for 1 minute, until cooked through. Drain, let cool, and peel.

Strain the soup into a bowl and transfer the solids to a food processor or blender. Puree until smooth, adding a little broth or water if necessary. Return the puree and the soup to the pot and whisk to combine.

Using the food processor, puree the roasted tomatoes, then strain through a coarse sieve (discard the solids). Stir the puree into the soup. Season with salt and pepper.

Ladle the soup in bowls and garnish each with a shrimp. Serve immediately.

烤蕃茄湯 *Roasted Tomato Soup* 209

# 煎沙甸魚配葡式辣醬
## Grilled Sardines over Portuguese Stew

我得承認我不是魚的忠實粉絲，但我卻從來不會對烤沙甸魚説不。朋友們都説我是個怪人，因為沙甸魚的氣味比其他魚類更腥，但這對我來説從來都不是問題。2018年3月，我去了葡萄牙波爾圖，當地盛產海鮮，我在當地品嚐了多款葡萄牙菜餚，包括這裡介紹的煎沙甸魚配葡式辣醬。因為這菜餚，更肯定了我對沙甸魚的喜愛。這辣汁跟沙甸魚是絕配，你也可以用它來搭配烤蝦或其他海鮮菜餚。

I need to confess that I am not a big fan of fish, but I can never say no to grilled sardines. My friends say I am really a weirdo because sardines tend to be very fishy and—some say—foul, but that's never been an issue for me. In March 2018, I visited the beautiful seaport of Porto and had a chance to sample many kinds of Portuguese dishes, including Grilled Sardines over Portuguese Stew. And it reaffirmed my affection for sardines. The stew is a good complement to the sardines, or you could also serve it with grilled shrimp or other seafood dishes.

材料（4人份量）
葡式辣醬

| | |
|---|---|
| 橄欖油 | 2湯匙 |
| 洋蔥切碎 | 120克 |
| 月桂葉 | 2塊 |
| 西班牙腸或葡萄牙腸切粒 | 200克 |
| 蒜頭去皮 | 12瓣 |
| 新鮮蕃茄切粒 | 120克 |
| 新薯切成4份 | 120克 |
| 新鮮百里香切碎 | 2茶匙 |
| 新鮮蕃茜切碎 | 2茶匙 |
| 新鮮羅勒切碎 | 2茶匙 |
| 雞湯 | 940毫升 |
| 鹽和胡椒粉 | 適量 |

Ingredient (Makes 4 servings)
Portuguese Stew

| | |
|---|---|
| Olive oil | 2 tablespoons |
| White onion, minced | 120g |
| Bay leaves | 2 |
| Chorizo or | |
| Portuguese sausage, diced | 200g |
| Garlic cloves, peeled | 12 |
| Fresh tomatoes, chopped | 120g |
| New potatoes, quartered | 120g |
| Fresh thyme, chopped | 2 teaspoons |
| Fresh parsley, chopped | 2 teaspoons |
| Fresh basil, chopped | 2 teaspoons |
| Chicken broth | 940ml |
| Salt and pepper | to taste |

| 沙甸魚 | | Sardines | |
|---|---|---|---|
| 新鮮或急凍沙甸魚 | 8條<br>（每條約80克） | Freesh or thawed<br>frozen sardines | 8 (about 80g each) |
| 橄欖油 | 2湯匙 | Olive oil | 2 teaspoons |

做法

葡式辣醬：把橄欖油加熱放入煎鍋裡以大火加熱。加入洋蔥及壓碎的月桂葉。炒8分鐘，直到洋蔥變軟。

加入肉腸，繼續煮2分鐘。加入蒜頭和蕃茄，再炒3至5分鐘，直到蕃茄變成糊狀及稍稍出水。

加入薯仔、百里香、蕃茜和羅勒。加入雞湯，用中火將汁料煮滾。再轉細火煮20分鐘，直至薯仔煮熟並軟身。用鹽和胡椒粉調味。

燒烤沙甸魚：將焗爐預熱至200°C。

將橄欖油灑在沙甸魚上。將沙甸魚放在烤架上，每邊烤2分鐘，直到魚皮呈微金黃色及酥脆。你也可以將魚放在有坑紋的煎鍋上用大火烤製，每邊煎4分鐘。

將葡式辣醬放在4個碟上，然後每碟放上兩條沙甸魚，上菜。

How to cook

For the stew: Heat the olive oil in a large saucepan over high heat. Add the onion and crush the bay leaves over them. Sauté for 8 minutes, until the onion is softened.

Add the sausage and continue to cook for 2 minutes. Add the garlic cloves and tomatoes and sauté for another 3 to 5 minutes, until the tomatoes turn mushy and slightly watery.

Stir in the potatoes, thyme, parsley, and basil. Add the broth and bring the liquid to a boil over medium heat. Reduce the heat to low and cook for another 20 minutes, until the potatoes are cooked and softened. Season with salt and pepper.

To grill the sardines: Heat the grill function of the oven to 200°C.

Toss the sardines with the olive oil. Place the sardines on the grill and cook for 2 minutes on each side, until the skin is lightly browned and crispy. You can also grill them on a ridged frying pan over high heat, cooking for 4 minutes on each side.

To serve, ladle the stew into 4 plates. Lay two sardines over each and serve.

# 葡萄牙蒜蓉雞
## Portuguese Garlic Chicken

我喜歡在冬天做這個菜餚。雖然製作需時，但只要將所有材料切好並倒入鑄鐵煎鍋中，這個簡單的步驟對我來說非常療癒，然後我期待聞到香味從鍋中滲出，我將整隻雞取出，放在烤架上將雞烤至金黃色。每當朋友們來我家吃晚飯，我總會把蒜蓉雞端到餐桌上，朋友們都吃得津津有味，伴著笑聲、音樂和葡萄酒。這就是我喜歡下廚的原因。

This is a dish that I love to cook during winter time. It takes a while but it's very therapeutic to chop all the ingredients and assemble them in a Dutch oven, then watch over the chicken, smelling the aroma from the pot, and finally take the bird out and brown it under the broiler. When friends come for supper and I bring garlic chicken to the table, everyone just digs in. Laughter, music, and wine. This is what I love about cooking.

| 材料（4人份量） | | Ingredient (Makes 4 servings) | |
|---|---|---|---|
| 洋蔥切薄片 | 1個 | White onion, thinly sliced | 1 |
| 蒜頭切薄片 | 6瓣 | Garlic cloves, thinly sliced | 6 |
| 新鮮蕃茄連皮切碎 | 2個 | Fresh vine tomatoes, chopped with skin on | 2 |
| 焗火腿切粒 | 150克 | Baked ham, chopped | 150g |
| 提子乾 | ⅓杯 | Raisins | ⅓ cup |
| 雞（走地雞） | 1隻（550至700克） | Whole chicken (free-range) | 1 (550 to 700g) |
| 砵酒 | 120毫升 | Tawny port wine | 120ml |
| 法式芥末 | 1湯匙 | Dijon mustard | 1 tablespoon |
| 茄膏 | 2湯匙 | Tomato paste | 2 tablespoons |
| 雞湯 | 240毫升 | Chicken broth | 240ml |
| 粟粉 | 1湯匙 | Cornstarch | 1 tablespoon |
| 凍水 | 1湯匙 | Cold water | 1 tablespoon |
| 紅或白酒醋 | 1湯匙 | Red or white wine vinegar | 1 tablespoon |
| 鹽和胡椒粉 | 適量 | Salt and pepper | to taste |

做法

在鑄鐵煎鍋或不黏砂鍋中，將洋蔥、蒜頭，蕃茄、火腿和提子乾混合在一起。從內到外沖洗雞隻然後抹乾。將雞隻放在拌好的洋蔥等食材上。

將砵酒、法式芥末、茄膏和雞湯放入碗中拌勻。淋在雞隻上。蓋上蓋子，用細火煮約2小時，直至能刺入大腿骨附近的肉並且肉質軟腍。

預熱烤架。將雞隻放烤盤的架子上，雞胸向上。將雞隻與烤爐放在離火10至15厘米上加熱，直到金黃色。不超過5分鐘。然後轉到溫暖的盤子保溫。

在一個小碗裡將粟粉和水一起拌勻。撇去鍋中汁液的脂肪。拌入粟粉水並調至大火加熱。蓋上蓋子，攪拌2至3次，直至汁液變濃，約5分鐘。加入醋，最後加入鹽和胡椒粉調味。

How to cook

In a cast iron Dutch oven or nonstick casserole, combine the onion, garlic, tomatoes, ham, and raisins. Rinse the chicken inside and out and dry. Place the chicken on top of the onion mixture.

Mix the port, mustard, tomato paste, and broth in a bowl. Pour over the chicken. Cover the pot and cook over low heat for about 2 hours, until the meat near a thighbone is tender when pierced.

Preheat the broiler. Transfer the chicken, breast side up, to the rack of a broiler pan. Broil 10 to 15cm from the heat until golden brown. This should take no more than 5 minutes. Transfer to a warm platter and keep warm.

Whisk together the cornstarch and water in a small bowl. Skim and discard any fat from cooking liquid in the pot. Stir in cornstarch mixture and increase the heat to high. Cover and cook, stirring 2 or 3 times, until the gravy thickens, about 5 minutes. Stir in the vinegar and season to taste with salt and pepper.

# 葡萄牙法式多士
## Rabanadas (Portuguese French Toast with Port)

葡萄牙法式多士Rabanadas是聖誕節期間其中一種傳統葡萄牙應節甜品。我記得在聖誕節子夜彌撒之後，我們總會準備一層層多士，上面灑滿了肉桂糖，爸爸媽媽喝著砵酒，而我和妹妹則喝著熱朱古力，大家邊飲邊吃，慶祝聖誕。這多士不用炸得太乾涸，凍或暖食皆可。

Rabanadas is one of the many traditional Portuguese desserts served at Christmas. I remember having a pile of the sweet fried bread topped with cinnamon sugar after midnight mass on Christmas Eve, with my parents enjoying their port wine while my sister and I had hot chocolate. It's very moist, rich slices of fried crusty bread, a bit like the pain perdu of French cuisine. It can be served either warm or cold.

| 材料（4人份量） | | Ingredient (Makes 4 servings) | |
|---|---|---|---|
| 糖漿 | | Syrup | |
| 白糖粉 | 300克 | Caster sugar | 300g |
| 檸檬皮 | 1隻檸檬 | Peeled lemon zest | from 1 lemon |
| 肉桂條 | 1條 | Cinnamon stick | 1 |
| 砵酒 | 2湯匙 | Tawny port wine | 2 tablespoons |
| | | | |
| 麵糊 | | Batter | |
| 牛奶 | 750毫升 | Milk | 750ml |
| 雞蛋發勻 | 2隻 | Eggs, beaten | 2 |
| 肉桂粉 | 1湯匙 | Ground cinnamon | 1 tablespoon |
| 白糖粉 | 1湯匙 | Caster sugar | 1 tablespoon |
| 雲呢拿油（可刪） | 1茶匙 | Vanilla extract (optional) | 1 teaspoon |

| | | | |
|---|---|---|---|
| 法式長棍麵包 | 切成3厘米厚，1條（隔夜) | Baguette, cut into 3cm-thick slices, dried overnight | 1 |
| 米糠油 | 1杯 | Rice bran oil | 1 cup |
| 白糖粉 | 250克 | Caster sugar | 250g |
| 肉桂粉 | 2茶匙 | Ground cinnamon | 2 teaspoons |
| 糖霜 | 3湯匙 | Icing sugar | 3 tablespoons |

做法

糖漿做法：將白糖粉、檸檬皮、肉桂條和砵酒倒入平底鍋中，用最細火煮5至10分鐘。只需要將糖漿微微加熱，當你能聞到檸檬和肉桂的香味，就可以把糖漿離火。

麵糊做法：將所有材料倒入一個大的淺碗裡。將麵包浸泡在麵糊中2至3分鐘，確保整片麵包都沾滿麵糊。

將煎鍋上的油以中火加熱。分批放入麵包片，每邊煎約1分鐘，直到金黃色。將煎好的麵包放在碟上的餐巾紙吸油及放涼

在淺碗中，用手指將白糖粉和肉桂粉均勻攪拌。逐片麵包放入碗中的肉桂糖然後撈出，再放到大的深盤中。

再次以中火加熱糖漿。將熱糖漿鋪滿麵包上，然後灑上糖霜。

How to cook

For the syrup: Put the caster sugar, lemon zest, cinnamon stick, and port in a saucepan and simmer over very low heat for 5 to 10 minutes. You only want to warm up the liquid—remove the saucepan from heat as soon as you can smell the aroma from the lemon and cinnamon.

For the batter: Combine all the ingredients in a large shallow bowl. Soak the bread in the batter for 2 to 3 minutes, making sure each side is coated and well covered.

Heat the oil on a frying pan over medium heat. In batches, add the bread slices and fry for about 1 minute on each side, until golden. Transfer to a paper towel–covered plate to drain and cool slightly.

In a shallow bowl, combine the caster sugar and cinnamon and mix well with your fingers. Dredge each bread slice in the cinnamon sugar and transfer to a large deep serving dish.

Reheat the syrup over medium heat. Be generous as you pour the hot syrup over the bread, then top with icing sugar.

 **Tips** 雲呢拿雪糕與這甜品是絕配，如果你想吃得健康點，可配雪葩。

Vanilla ice cream goes very well with this dessert, or a frozen sorbet if you want to sound slightly health conscious.

# 蝦多士
## Shrimp Toast

蝦多士是一種中國點心，在一塊小三角形麵包上鋪上蝦膠和馬蹄，掃上蛋漿，然後烤焗或油炸而成。我的家傳食譜沒有用上馬蹄，也沒有掃上蛋漿，我們喜歡用蝦多士點李派林喼汁食用。

Shrimp toast or prawn toast（蝦多士）is a Chinese dim sum dish of small triangles of bread that are brushed with egg and coated with minced shrimp and water chestnuts, then cooked by baking or deep-frying. My family's version doesn't use water chestnuts, egg is not brushed on, and we love to dip the shrimp toast in Worcestershire sauce.

| 材料（4人份量） | | Ingredient (Makes 4 servings) | |
| --- | --- | --- | --- |
| 鮮蝦 | 300克 | Fresh shrimp | 300g |
| 免治豬肉 | 50克 | Minced pork | 50g |
| 鹽和胡椒粉 | 適量 | Salt and pepper | to taste |
| 雞蛋輕輕打發 | 1隻 | Egg, lightly beaten | 1 |
| 粟粉 | 1茶匙 | Cornstarch | 1 teaspoon |
| 三明治厚切白麵包 | 4片 | White sandwich bread | 4 slices, thick |
| 橄欖油或米糠油 | 煎炸用 | Olive oil or rice bran oil | for frying |
| 李派林喼汁 | 拌多士用 | Worcestershire sauce | for dipping |

做法

將蝦去殼、去腸、洗淨，然後剁碎。（如果蝦身細小，可以不用剁碎）將蝦和豬肉放在大碗裡，用鹽和胡椒粉調味。醃15分鐘。

將蛋漿和粟粉拌勻。倒入碗中的蝦和豬肉並再次拌勻。

將麵包對角一切，切成4塊三角形，並將蝦膠均勻地塗在每片三角形麵包上。

在平底鍋或煎鍋中將6厘米的油加熱，直到你看到微微煙霧從油中冒出。多士放在罩籬上，蝦膠向下，每次放四件多士落油炸1分鐘左右。反轉多士並繼續炸數秒鐘，直至多士變成金黃色。將多士從油取出，放於餐巾紙上瀝乾多餘油份。準備一小碗李派林喼汁，趁熱把多士沾點喼汁食用。

How to cook

Peel, devein, and wash the shrimp, then chop. (You can skip the chopping if the shrimp are very small.) Combine the shrimp and pork in a large bowl and season with salt and pepper. Let sit for 15 minutes.

Whisk together the beaten egg and cornstarch. Add to the bowl and mix well with the shrimp and pork.

Cut each slice of bread into 4 triangles and spread the mixture evenly over each.

Heat 6cm oil in a saucepan or deep skillet until you can barely see some smoke coming out from the oil. In batches of 4, place the toasts, shrimp-side down, on a slotted spoon and carefully lower into the oil. Fry for about 1 minute. Turn each piece and continue frying for a few seconds more, until the bread is golden. Remove from the oil with a skimmer and drain on paper towels. Serve hot with small bowls of Worcestershire sauce for dipping.

蝦多士 *Shrimp Toast* 225

# 香橙豬肘骨
## Cotovelo de Porco (Pig Elbow Bone)

以豬為食材的葡國料理有超過100種，這道香橙豬肘骨是一道已經失傳的懷舊食譜。小時候，嫲嫲很喜歡煮這道菜給我吃。

There must be more than 100 Portuguese recipes for porco (pork). This is an authentic Macanese dish that I used to have when living with Avo.

| 材料（4人份量） | |
|---|---|
| 新鮮豬肘骨 | 4至8件 |
| 白酒 | 3湯匙 |
| 豉油 | 1湯匙 |
| 鹽 | ½湯匙 |
| 白胡椒粉 | 1茶匙 |
| 橄欖油 | 4湯匙 |
| 乾蔥頭切片 | 1粒 |
| 蒜頭切碎 | 3瓣 |
| 月桂葉 | 2塊 |
| 西班牙腸或葡萄牙腸切粒 | 1條 |
| 雞湯 | 4湯匙 |
| 橙皮蓉 | 1湯匙 |
| 鮮橙汁 | 3湯匙 |
| 鹽和胡椒粉 | 適量 |

| Ingredient (Makes 4 servings) | |
|---|---|
| Fresh pig elbow bone | 4 to 8 pieces |
| Dry white wine | 3 tablespoons |
| Soy sauce | 1 tablespoon |
| Salt | ½ tablespoon |
| Ground white pepper | 1 teaspoon |
| Olive oil | 4 tablespoons |
| Shallot, sliced | 1 |
| Garlic cloves, minced | 3 |
| Bay leaves | 2 |
| Chorizo or | |
| Portuguese sausage, diced | 1 |
| Chicken broth | 4 tablespoons |
| Grated orange zest | 1 tablespoon |
| Fresh orange juice | 3 tablespoons |
| Salt and pepper | to taste |

做法

用水沖洗豬肘骨，並用酒、豉油、鹽和胡椒粉搓勻整塊豬肘骨。封好並放入雪櫃冷藏至少2小時。

用炒鑊或煎鍋以大火加熱2湯匙橄欖油。落豬肘骨，煎約10分鐘，每邊5分鐘，直到豬肘骨轉為微微金黃色。然後上碟並放在一邊待用。

用剛煮過豬肘骨的炒鍋或煎鍋，以中火加熱剩餘的2湯匙橄欖油。加入乾蔥頭、蒜頭、月桂葉和肉腸，煮約5分鐘，直到乾蔥軟身，及油份從肉腸滲出。

將豬肘骨放回鍋中再煮2分鐘。加入雞湯，蓋上蓋子，用中低火煮10分鐘，直到豬肘煮熟為止。

離火，加入橙皮蓉和橙汁，拌勻。用鹽和胡椒粉調味即可上碟。

How to cook

Rinse the pig elbow bone and rub with the wine, soy sauce, salt and pepper. Cover and refrigerate for at least 2 hours.

Heat 2 tablespoons of the olive oil in a wok or frying pan over high heat. Add the pig elbow bone and cook for about 10 minutes, 5 minutes on each side, until the meat is lightly browned. Transfer to a plate and set aside.

Using the same wok or frying pan, heat the remaining 2 tablespoons olive oil over medium heat. Add the shallot, garlic, bay leaves, and chorizo and cook for about 5 minutes, until the shallot is soft and oil from the sausage releases.

Add the pork back to the wok and cook for another 2 minutes. Pour over the broth, cover, and cook over medium-low heat for 10 minutes, until the pork is cooked through.

Remove from the heat stir in the orange zest and juice. Season with salt and pepper and serve.

明盡
乃至無

提行識無眼耳鼻舌身意無色
法無眼界乃至無意識界
是故空中無色無
味觸

聖宅...瞻礼化好，要宫
瞻礼的特点。爸爸买...
也吃水果，爸爸买水果，要
同种题的 做一个自修的生果

所從 嫲嫲離開了
，再加上很多 aunties
都相繼離開

的交... 晚
了 十...
xxx園

# 蠔油青檸煮雞
## Chicken with Oyster Sauce and Lime

難道我是一名雞癡？如果你看過我之前的故事「開竅了」（p.51），你便知道我對食物的熱愛始於一塊簡單的炸雞，我在那故事裡介紹了三道以雞為主要食材的菜餚。現在提供另一款以雞為題的食譜。這是我一位親戚的懷舊食譜。你能想像蠔油和青檸汁結合的味道嗎？

Do you think I am a "chickenoholic"? If you read my earlier story, "Awakened" (p 51), you learned that my love for food started with a simple piece of fried chicken, and then I included three of my favorite chicken dishes. And now, here is another. It's an old recipe from one of my relatives. Can you imagine the taste of the combination of oyster sauce and lime juice? No idea? Cook it and let me know what you think.

材料（4人份量）

| | |
|---|---|
| 雞（走地雞） | 1隻（550至700克） |
| 鹽 | 1.5湯匙 |
| 白酒 | 4湯匙 |
| 橄欖油 | 6湯匙 |
| 蒜頭切碎 | 6瓣 |
| 乾蔥切碎 | 6粒 |
| 薑 | 6片 |
| 青檸汁 | 3湯匙 |
| 蠔油 | 2湯匙 |
| 雞湯或水 | 4湯匙 |
| 蔥和芫茜切碎 | 裝飾用 |

Ingredient (Makes 4 servings)

| | |
|---|---|
| Whole chicken (free-range) | 1 (550 to 700g) |
| Salt | 1 ½ tablespoons |
| Dry white wine | 4 tablespoons |
| Olive oil | 6 tablespoons |
| Garlic cloves, finely chopped | 6 |
| Shallots, finely chopped | 6 |
| Fresh ginger | 6 slices |
| Lime juice | 3 tablespoons |
| Oyster sauce | 2 tablespoons |
| Chicken broth or water | 4 tablespoons |
| Chopped scallion and cilantro | for garnish |

做法

用水沖洗雞隻並用1湯匙鹽搓勻整隻雞的內外面。讓雞待乾30至45分鐘。

用剩下的½湯匙鹽和白酒搓勻於雞隻上。封好雞隻並醃30分鐘。

將雞切成8至10塊。加橄欖油於煎鍋中，用中火加熱。加入蒜頭、乾蔥頭和薑，炒2分鐘直至釋出香味，然後加入雞件煮約5分鐘。加入青檸汁和蠔油再煮2分鐘。

最後加入雞湯，蓋上蓋子，用細火煮10分鐘，直至雞肉煮熟並嫩脧。加一點切碎的蔥花和芫茜於餸面作裝飾。

How to cook

Rinse the chicken and rub inside and out with 1 tablespoon of the salt. Leave to dry for 30 to 45 minutes.

Rub the chicken with the remaining ½ tablespoon salt and the wine. Cover and marinate for 30 minutes.

Cut the chicken into 8 to 10 pieces. Heat the olive oil in a frying pan over medium heat. Add the garlic, shallots, and ginger and sauté for 2 minutes to release the aroma, then add the chicken and cook for about 5 minutes. Add the lime juice and oyster sauce and cook for another 2 minutes.

Add the broth, cover, and cook over low heat for 10 minutes, until the chicken is cooked through and tender. It's good to serve the chicken with a bit of chopped scallion and cilantro.

蠔油青檸煮雞 *Chicken with Oyster Sauce and Lime*　　233

# 葡式甜飯
## Creamy Portuguese Rice Pudding (Arroz Doce)

Arroz Doce（甜飯的意思）是其中一道我吃了會頓時覺得非常窩心的菜式, 我稱它為comfort food。如果你在互聯網上搜索，你會發現Arroz Doce被列為葡萄牙十大甜品之一。葡萄牙人喜愛它是因為這甜品味道濃郁而且軟綿綿，為葡萄牙人帶來了許多童年甜蜜的回憶。

Arroz Doce (literally, "sweet rice") is one of my comfort foods. If you search on the internet, you will find that it is often listed as one of the top ten desserts of Portugal. I'm pretty sure it is loved by the Portuguese because the rich, creamy pudding brings with it a lot of childhood memories.

| 材料（4人份量） | | Ingredient (Makes 4 servings) | |
|---|---|---|---|
| 白米 | 50克 | Long-grain rice | 50g |
| 牛奶 | 960毫升 | Milk | 960ml |
| 無鹽牛油 | 100克 | Unsalted butter | 100g |
| 雲呢拿油 | 1茶匙 | Vanilla extract | 1 teaspoon |
| 白糖粉 | 250克 | Caster sugar | 250g |
| 檸檬皮蓉 | 5湯匙<br>（大約2個檸檬） | Grated lemon zest | 5 tablespoons<br>(from 2 lemons) |
| 肉桂條 | 1條 | Cinnamon stick | 1 |
| 蛋黃 | 5隻 | Egg yolks | 5 |
| 鹽 | 一少撮 | Salt | small pinch |
| 肉桂粉 | 裝飾用 | Ground cinnamon | for garnish |

做法

將米和大量的水放入平底鍋中，蓋上蓋子。用大火煮8分鐘。然後將飯用篩子瀝乾。

把飯放回鍋裡。加入牛奶、牛油、雲呢拿油和糖。用中火將材料煮至起泡，然後加入檸檬皮和肉桂條。蓋上蓋子，用細火煮飯，15至20分鐘左右，直到米飯煮至你喜歡的質感。

加入蛋黃和少許鹽。均勻攪拌，米飯熱力也會同時把蛋黃煮熟。取出肉桂條，將布甸倒入淺盤中。放上肉桂粉和多一點檸檬皮作裝飾。此款甜品可熱食或凍食。

How to cook

Place the rice in a saucepan and cover with lots of water. Bring to a boil over high heat, cover and cook for 8 minutes. Drain in a sieve.

Return the rice to the saucepan. Add the milk, butter, vanilla, and sugar. Cook over medium heat until the mixture first begins to bubble, then add the lemon zest and cinnamon stick. Cover and let the rice simmer over low heat for 15 to 20 minutes, until the rice is cooked to your preferred texture.

Stir in the egg yolks and a small pinch of salt. Stir well while the heat of the rice cooks the egg yolks. Remove the cinnamon stick and pour the pudding into a shallow dish. Garnish with ground cinnamon and more lemon zest. Serve hot or cold.

母親與我

Mother and Son

*以下文章是在2018年2月13日在我的blog發表，
當天為我首次在Towngas Cooking Centre 舉行料理班。

# 更上一層樓*

在我做blogger的最初階段，我都沒有告訴家人和身邊的朋友，只是有一次妹妹無意中在食譜平台Cook1Cook看見一幅似曾相識的相片，那碗洋蔥湯，那張枱布……這一年來，家人都默默支持我的食之旅，追蹤我的行蹤。而媽媽要make sure 我沒有把一些家族食譜改了過來，因為她知道我很喜歡make a change。LOL！！！

上星期媽媽在WhatsApp留了一段voice message給我，問我可曾記得小時候我們往銅鑼灣，嫲嫲總喜歡叫司機把車停在禮頓中心，然後走過當時位於禮頓中心地庫的Towngas，因為很多時剛會碰到在上料理班的Auntie Souza。

Well，我當然記得啊！ 還記得在Towngas的隔鄰好像有一兩間時裝店，我最喜歡就是嫲嫲在那裡shopping，而我就走到旁邊的cooking centre，在玻璃窗外看著在廚藝學校煮餸的aunties and uncles，好想走入去玩。每次回到家便嚷著嫲嫲到kitchen煮東西，我在旁看，就是這樣的，我從嫲嫲的身上學了很多的食譜。

記得去年我曾經對自己說過如果有一天能夠在Towngas開料理班就好了。這一年來原來已經不知不覺的開了超過100場料理班，而跟合作的廚藝學院由最初的一間到目前的五間，so far so good，非常smooth 的合作關係。雖然有些時候都會頗為 exhausting "mentally"。但我始終認為有機會就要把握，不要放棄。所以一直的去「拚」。

有很多粉絲知道我將會在Towngas新一季的課程任教之後，都替我開心，給了我很多的祝福語，有一位更對我說是「更上一層樓」。我當然開心，但對我來說更有一層更深的意義。因為這個 brand name 某程度上是給了我很多童年的回憶，而我對廚藝的興趣都是從嫲嫲和媽媽開始。

今天2月13日是我首場在Towngas Cooking Centre的料理班，教班的方式和風格不會有甚麼不同。就是跟每位同學開開心心的分享，做自己喜歡做的事，一齊講，一齊笑。This is the joy of cooking。

原來媽媽也拿了2本煤氣烹飪中心的課程小冊子，一本給了我，一本給爸爸。

我對自己說：Yes，I made it。

*an adaption of my blog
post of February 13, 2018

# A New Chapter *

Today, I received a WhatsApp voice message from mom, asking if I remember the days when Avo took me to Leighton Centre. Of course I remember: The driver pulled right in front of Leighton Centre, where there was a staircase in the middle directly leading to the Towngas Cooking Centre. Avo would spend an hour shopping at the cozy fashion store next to Towngas. But I was bored with her shopping (obviously), so ended up watching the cooking classes. That's what led me, back at home, to ask Avo to play the game of cooking class with me, and that's how I learned so many dishes from her.

I told myself last year that if someday I could host a cooking class at Towngas, that would be awesome. I worked harder and harder, gradually expanding my job experience by teaching at five different cooking schools, and ended up hosting over 100 classes in just a year.

After I'd been teaching for a number of months at different cooking schools, a few of my clients recommended me to the Towngas Cooking Centre. The school checked out my social media and asked me if I'd like to join their staff. I was thrilled to be invited as it's the most reputable cooking school in Hong Kong.

When clients and fans heard that I was to teach at Towngas, I received a lot of messages and greetings congratulating me. I'm thrilled to hear those, of course, but for me, having the opportunity to teach at Towngas has a much deeper meaning. It brings me full circle, bringing up a lot of childhood memories, especially my special relationship with Avo.

So, today, just one year after my very first cooking class, is my first cooking class at Towngas and I am all for it. As is my mom—she took two program booklets with my photo and class info from Towngas, one for me, one for papa.

I made it!

第 三 部

# 明日

明日？會想的，但不會時時刻刻去想。
隨心的想，隨心的做。信自己的感覺吧！

# *Tomorrow*

I don't think too much when it comes to uncertainties.
I just follow my feeling and do whatever I consider it's right.

# 黑色記事簿

我不時都需要翻看我的黑色記事簿，裡面記錄了我的食譜和做菜時我寫下的所有測試筆記。我在「鹹蝦燦食之旅」這個project開始時便養成了寫烹飪日誌的習慣，記錄了我食譜中各種食材、份量和步驟的變化和調整。你可能無法相信，我把Batatada焗蕃薯糕（p.83）的配方調整超過10次，才做出恰到好處的口感和美味。

自2017年開始的「鹹蝦燦食之旅」，我曾向自己承諾在Facebook和Instagram分享100道菜餚。2017年12月27日，炸雞肉餅成為我的第100道菜。I made it！在這100道菜當中，近三分之二是葡萄牙菜和澳葡土生菜。其中有一些更是我自己的創作菜式，我稱之為Neo Macanese新派澳葡菜。

為什麼要花時間創造自己的菜式？因為我想將葡國菜提升到一個新的水平，開啟新的一頁。傳統的土生澳葡菜是一代傳一代，所以西洋一品鍋Daibo（p.113）總是西洋一品鍋Daibo。我會想：把香茅加入一品鍋會如何？或者將魚露加入在我任何一個estufada的菜餚中？這是我非常感興趣的事情，特別是將泰國菜與葡萄牙菜融為新菜式。

有一次，我在做葡萄牙鴨飯。我依照傳統方式烹調湯底和切碎鴨肉，但我沒有按照媽媽給我的食譜把飯焗至脆口，也沒有將鴨肉放在米飯之間，我用上新的烹調方式，我用意大利米燴成飯，然後加了橙皮和葡萄牙肉腸。這配方是經過多次測試、筆錄以及調整才創出新的美味，並記錄在黑色記事簿中。這就是烹飪的樂趣，the joy of cooking。

# The Black Book

From time to time, I need to go back to my Black Book, which is filled with my recipes and all the notes I've scribbled down after cooking and testing. I started this cooking journal at the beginning of my Food Journey project to document the changes and tweaks to the measurements, ingredients, and steps of my recipes. Believe it or not, I have changed the Batatada recipe (p.83) more than ten times in order to get just the right texture and flavor.

With the Food Journey project, started in 2017, I promised myself that I'd cook 100 dishes and feature each one on Facebook and Instagram. On December 27, 2017, chicken croquette became my 100th dish. I made it! Of all those 100 dishes, almost two-third are from Portuguese and Macanese cooking. But some of them are my own creative cooking, which I like to call Neo Macanese.

Why spend the time to create something on my own? It's simply that I want to take Macanese cooking to a new level, to open a new chapter. Most of the dishes that we know and that I teach are family recipes that have been passed from one generation to the next. So Daibo (p.113) is always Daibo. But I would wonder: How about adding lemongrass to Daibo? Or using fish sauce in one of my estufada dishes? This is something that I was very interested into pursuing, especially blending Thai cooking with Portuguese cooking.

One day, I studied Portuguese duck rice. I followed the traditional way of making a broth and shredding the duck meat. But instead of baking the rice to give it a crunchy texture on top with the duck meat hidden between layers of rice, I created a Neo Duck Rice by using the risotto technique for cooking the rice and then seasoning it with orange zest in addition to Portuguese sausage. It took a lot of testing and notes and changes to the recipe in the Black Book to get it right. But this is simply the joy of cooking.

# 葡萄牙鴨飯
## Portuguese Duck Rice

Arroz de pato（葡萄牙鴨飯）是傳統的農家菜。就像中國人喜歡在冬天將雞肉和飯放在熱騰騰的瓦煲一樣，葡萄牙人也非常喜歡他們的鴨飯。當你在澳門旅遊或在香港的葡萄牙餐廳用餐時，它是一款常見的菜餚。我的家傳食譜與傳統的略有不同，因為我會用上鴨腿而不是整隻鴨子。

This is arroz de pato, the rustic, comforting traditional Portuguese dish. Just as the Chinese love their chicken and rice cooked in hot clay pot during winter, the Portuguese love their duck rice. This is also a dish that is commonly found when visiting Macau or dining in Portuguese restaurants in Hong Kong. My family recipe is slightly different from the traditional because it uses duck thighs instead of whole duck.

| 材料（4人份量） | | Ingredient (Makes 4 servings) | |
|---|---|---|---|
| 鴨腿 | 2隻（共約450克） | Duck thighs | 2 (about 450g total) |
| 鹽 | 1茶匙 | Salt | 1 teaspoon |
| 黑胡椒粉 | 1茶匙 | Black pepper | 1 teaspoon |
| 雞湯 | 946毫升 | Chicken broth | 946ml |
| 水 | 300毫升 | Water | 300ml |
| 洋蔥去皮 | 2個 | White onions, peeled | 2 |
| 蒜頭去皮 | 4瓣 | Garlic cloves, peeled | 4 |
| 月桂葉 | 2片 | Bay leaves | 2 |
| 橄欖油（可以不加） | 1湯匙 | Olive oil (optional) | 1 tablespoon |
| Agulha長米 | 250克 | Agulha (long-grain) rice | 250g |
| 西班牙腸或葡萄牙腸，切成3毫米厚 | 50克 | Chorizo or Portuguese sausage, sliced 3mm thick | 50g |
| 蛋黃 | 2湯匙 | Egg yolk | 2 tablespoons |

## 做法

用鹽和胡椒粉將鴨腿調味，封好，冷藏2小時。

將雞湯和水倒入大煎鍋或煲中加熱。加入整個洋蔥、蒜頭和月桂葉。煮滾後，調至中火煮15分鐘。

用中火加熱中型煎鍋，直至煎鍋變熱。（鴨有很多脂肪，所以我不會落油。如你喜歡的話，可以加入1湯匙橄欖油來煎鴨腿）。將鴨腿放落煎鍋並煎至微金黃色，每邊煎約5至8分鐘。

從煎鍋中取出鴨腿和油，放入雞湯中，用細火煮60至90分鐘。當鴨肉變臉並開始離骨時，就代表鴨湯煮好了。

將鴨腿轉到碟上。並放涼一會，當它不再燙手時，就可以去皮、去骨，將鴨腿切成一口大小的鴨片，放在一旁。從湯中取出洋蔥、蒜頭和月桂葉並丟棄。

將焗爐預熱至200°C。

將720毫升鴨湯放入中型平底鍋中煮滾。洗米，然後加入沸騰的鴨湯中。用中火煮約15分鐘，直到米飯半熟。

將一半米飯放入寬而淺的陶瓷煲或不黏鍋的底部。放上切碎的鴨肉和幾片肉腸，然後蓋上另一層米飯。

用蛋黃刷上飯面（這樣有助於形成脆皮）然後在飯面放上剩餘的肉腸切片。焗10分鐘，直到頂部變成金黃色和略微鬆脆，上碟。

## How to cook

Season the duck thighs with the salt and pepper, cover, and refrigerate for 2 hours.

In a large saucepan or pot, combine the broth and water. Add the whole onions, garlic, and bay leaves. Bring to a boil, reduce the heat, and simmer over medium heat for 15 minutes.

Heat a medium skillet over medium heat until hot. (The duck should give off a lot of fat, so I tend to skip using oil. But you can add 1 tablespoon olive oil to sear the duck thighs if you like.) Add the thighs and sear until lightly browned, 5 to 8 minutes on each side.

Add the duck thighs and oil from the skillet to the broth and simmer over low heat for 60 to 90 minutes. When the duck meat turns tender and begins to fall off the bones, it's ready, and the duck broth is done.

Transfer the duck thighs to a plate. When they're cool enough to handle, remove the skin, pull the meat from the bones, and shred into bite-sized pieces. Set aside. Remove the onions, garlic, and bay leaves from the broth and discard.

Preheat the oven to 200°C.

Bring 720ml of the duck broth to a boil in a medium saucepan. Rinse the rice and add to the boiling broth. Simmer over medium heat for about 15 minutes, until the rice is halfway done.

Transfer half of the rice to the bottom of a wide, shallow earthenware pot or nonstick casserole. Cover with the shredded duck meat and a few slices of chorizo, then another layer of rice.

Brush the top with egg yolk (this will help form the crust) and top with the remaining slices of chorizo. Bake for 10 minutes, until the top is browned and slightly crunchy. Serve.

你可以將剩餘的鴨湯放在雪櫃中長達3週。我喜歡用鴨湯炆燉菜餚，或者用一些切碎的鴨肉、洋蔥、西芹和紅蘿蔔做湯。

You can keep leftover duck broth in the freezer up to 3 weeks. I normally like to use duck broth for making stew, or a soup with some shredded duck meat, onions, celery, and carrots.

# 新派手拆鴨肉飯
## Neo Duck Rice

葡萄牙鴨飯（p.249）是冬季佳餚：我這個版本用上意大利米，是一款全年皆宜的菜式。鴨肉與橙很匹配，所以我於這個菜式加了橙皮和橙汁，並用Arborio意大利米代替長米。煮法與葡萄牙鴨飯非常相似，這個鴨飯比傳統做法清淡，更能提升食材味道。兩個版本我都喜歡。

Portuguese Duck Rice (p.249) is good for winter; my version of risotto-style duck rice is good all year-round. Duck meat goes well with orange, so I added orange zest and orange juice, and use Arborio rice instead of long-grain rice. The method is quite similar to Portuguese Duck Rice, but with a lighter, brighter flavor. I love both versions.

| 材料（4人份量） | | Ingredient (Makes 4 servings) | |
|---|---|---|---|
| 鴨腿 | 2隻（共約450克） | Duck thighs | 2 (about 450g total) |
| 鹽 | 1茶匙 | Salt | 1 teaspoon |
| 黑胡椒粉 | 1茶匙 | Black pepper | 1 teaspoon |
| 橄欖油（可以不加） | 1湯匙 | Olive oil (optional) | 1 tablespoon |
| 雞湯 | 946毫升 | Chicken broth | 946ml |
| 水 | 300毫升 | Water | 300ml |
| 洋蔥去皮 | 2個 | White onions, peeled | 2 |
| 蒜頭去皮 | 4瓣 | Garlic cloves, peeled | 4 |
| 月桂葉 | 2塊 | Bay leaves | 2 |
| 洋蔥切碎 | ½個 | Onion, minced | ½ |
| Arborio（意大利米） | 250克 | Arborio (risotto) rice | 250g |
| 白酒 | 50毫升 | Dry white wine | 50ml |
| 西班牙腸或葡萄牙腸切粒 | 50克 | Chorizo or Portuguese sausage, diced | 50g |
| 橙汁 | 5湯匙 | Orange juice | 5 tablespoons |
| 無鹽牛油 | ½湯匙 | Unsalted butter | ½ tablespoon |
| 橙皮 | 1湯匙 | Grated orange zest | 1 tablespoon |
| 新鮮蕃茜切碎 | 2茶匙 | Fresh parsley, minced | 2 teaspoons |

做法

用鹽和胡椒粉將鴨腿調味，封好，冷藏2小時。

用中火加熱中型煎鍋，直至煎鍋變熱。（鴨有很多脂肪，所以我不會落油。如你喜歡的話，可以加入1湯匙橄欖油來煎鴨腿）。將鴨腿放落煎鍋並煎至微金黃色，每邊煎約5至8分鐘。將鴨腿放在一旁，煎鍋及鍋中的油稍後待用。

將雞湯和水倒入大煎鍋或煲中加熱。加入2個洋蔥、蒜頭和月桂葉。煮滾後，調至中火煮15分鐘。

將鴨腿放入雞湯中，用細火煮60至90分鐘。當鴨肉變腍並開始離骨時，就代表鴨湯煮好了。

將鴨腿轉到碟上，放涼一會，當它不再燙手時，就可以去皮、去骨，將鴨腿用手拆為一口大小的鴨肉，放在一旁。從湯中取出洋蔥、蒜頭和月桂葉並丟棄。

準備煎鍋：將剁碎的洋蔥加入煎鍋內的剩餘鴨油中，用中火加熱，偶爾攪拌，直至軟身，約5分鐘。加入意大利米，與鴨油拌勻，煮至米的週邊透明，約2分鐘。灑進白酒，煮至米飯把汁料完全吸收。加入熱鴨湯，邊煮邊攪拌，直到湯液近乎蒸發。繼續加入鴨湯，攪拌，重複以上步驟，煮大約15分鐘。

加入肉腸粒，接著鴨肉碎。重複加入鴨湯、攪拌，直到飯堆稍微凹陷，並且飯變嫩，而飯堆中間有微微阻力，需要5到10分鐘或更長。拌入橙汁和牛油，加入鹽和胡椒粉調味。

上碟時，撒上橙皮和蕃茜，然後馬上享用。

Season the duck thighs with the salt and pepper, cover, and refrigerate for 2 hours.

Heat a skillet over medium heat until hot. (The duck should give off a lot of fat, so I tend to skip using oil. But you can add 1 tablespoon olive oil to sear the duck thighs if you like.) Add the thighs and sear until lightly browned, 5 to 8 minutes on each side. Set the duck thighs aside, and keep the skillet and oil for later use.

In a large saucepan or pot, combine the broth and water. Add the 2 whole onions, the garlic, and bay leaves. Bring to a boil, reduce the heat, and simmer over medium heat for 15 minutes.

Add the duck thighs to the broth and simmer over low heat for 60 to 90 minutes. When the duck meat turns tender and begins to fall off the bones, it's ready, and the duck stock is done.

Transfer the duck thighs to a plate. When they're cool enough to handle, remove the skin, pull the meat from the bones, and shred into bite-sized pieces. Set aside. Remove the onions, garlic, and bay leaves from the broth and discard.

Back to the skillet: Add the minced onion to the fat and sauté over medium heat, stirring occasionally, until softened, about 5 minutes. Stir in the rice to coat with fat, and cook until translucent around the edges, about 2 minutes. Splash in the wine and continue cooking until it is absorbed. Add a ladleful of hot duck stock and cook, stirring continuously, until the liquid has almost burbled away. Keep up the rhythm of adding stock, stirring, and cooking for 15 minutes.

Stir in the diced chorizo, then the shredded duck. Resume ladling, stirring, and cooking until the rice slumps gently when mounded and is tender, but offers just the slightest bit of resistance in the middle, take 5 to 10 minutes or even longer. Stir in the orange juice and butter and season with salt and pepper to taste.

To serve, sprinkle with orange zest and parsley and rush to the table. Enjoy!

你可以將剩餘的鴨湯放在雪櫃中長達3週。我喜歡用鴨湯炆燉菜餚，或者用一些切碎的鴨肉、洋蔥、西芹和紅蘿蔔做湯。

You can keep leftover duck broth in the freezer up to 3 weeks. I normally like to use duck broth for making stew, or a soup with some shredded duck meat, onions, celery, and carrots.

# 我的同學，我的粉絲們

作為導師或KOL，其中一個關鍵是與同學和粉絲保持緊密而融洽的關係。我沒有花錢在社交媒體上「買粉絲」，我會把握每個在線上與粉絲互動的機會。在課堂上，我不會只留在自己的煮食台，看著學生自己做菜。我喜歡走到每位同學旁邊，逐一和他們交談。分享烹飪心得，這就是我所說的分享。儘管每次上完堂我都有點drain out，但我仍然是非常享受跟每位同學互動的機會。

我也盡量透過Facebook Live臉書直播功能與我的粉絲聊天，有空的時候我還會跟他們示範一些簡單的菜餚，如馬介休薯仔沙律配墨西哥粟米片。2018年3月，我帶粉絲到葡萄牙波爾圖，在那裡我們享受了一星期的美食之旅。眼見為實，我們品嚐了很多在課堂上教過的菜餚：薯蓉青菜湯（p.97），馬介休薯球（p.143）和傳統葡式蛋白餅Molotoff，

這是一種用蛋白和糖製成的清淡甜點，既滑溜，甜度又恰到好處。2018年9月，我們去到馬來西亞的馬六甲，品嚐了極具葡萄牙和馬拉混合菜元素的Kristang料理；2018年10月我跟12位同學和粉絲們去到泰國曼谷進行美食交流，還在曼谷著名的Blue Elephant烹飪學校上課。

我把粉絲看成為我的好友，他們也同樣把我當成好友。我陶醉於跟他們交流的時刻，特別是他們在上課時品嚐菜餚的那一個moment。我的美食之旅從來都沒有孤獨感，因為我總是給志趣相投的粉絲和朋友包圍著。

烹飪絕不是平凡乏味的事情。隨心的煮，用心的跟別人分享，讓烹飪變得更有趣味。

# My fans, My friends

One of the key challenges as an instructor or KOL (key opinion leader) is building a good rapport with my students and fans. I don't pay to recruit followers on social media, but I do take every opportunity to interact with my followers on the internet. And during classes, I don't stay at my own cooking station and watch students cook on their own. I prefer to walk around and talk to them one on one. This is what I call sharing.

And I try my best to make use of Facebook Live events to interact with my fans, chatting with them or sometimes showing them a simple dish like Potato Salad with Bacalhau and Tortilla Chips. In March 2018, I took my fans to Porto where we had a lovely week of gastronomy indulgence. Seeing is believing. We sampled a lot of dishes that I have taught in my classes: Caldo Verde (p.97), Bacalhau Fritters (p.143), and traditional Portuguese Molotoff, a light dessert made with egg whites and sugar that is the perfect amount of smooth and sweet. And we also traveled to Malacca, Malaysia, in September 2018 and Bangkok, Thailand, in October 2018 for gastronomy exchanges, including cooking lessons at Bangkok's renowned Blue Elephant Cooking School.

Some people might ask why I do all this and keep myself so busy? Well it's simply because I love cooking, and cooking is all about sharing.

I treat my fans as my friends and they have also all given back to me. I enjoy the interaction with them, and the moments when they savor the dish at the end of class. It's a moment of satisfaction and happiness that I look forward to in every class. My food journey has never been a lonely one as it's always surrounded by fans and friends alike.

Cooking is never something mundane. Make it simple, do it with love, and make it fun.

# 馬介休炸蝦丸
## Bacalhau Shrimp Balls

馬介休（鹹鱈魚）可以做出多種不同的菜式，而馬介休薯球（p.143）是其中最經典的。這款馬介休炸蝦丸的靈感來自炸蝦球（炸蝦棗），是一種傳統的潮州菜式，將蝦仁剁成蓉，混合肥豬肉和馬蹄，最後在滾油中炸至金黃色。我喜歡混合不同的食材和調味品來創造自己的食譜，我非常鼓勵大家也試一試。這也是一件非常療癒的事情。

There are many different ways to cook bacalhau (salted cod fish) and Bacalhau Fritters (p.143) is the most classic dish. These bacalhau balls were inspired by deep-fried shrimp balls (炸蝦棗), a traditional Chaozhou dish of shrimp chopped into a paste, mixed with fatty pork and water chestnuts, and finally deep-fried in very hot oil until golden brown. I enjoy creating my own recipes by mixing and matching different produce and condiments, and I strongly encourage everyone to do so. It's very therapeutic.

| 材料（8人份量） | | Ingredient (Makes 8 servings) | |
|---|---|---|---|
| 急凍鱈魚柳 | 450至500克 | Cod fish fillets (frozen) | 450 to 500g |
| 海鹽 | 1湯匙 | Sea salt | 1 tablespoon |
| 蝦煮熟並切碎 | 200克 | Shrimp, cooked and chopped | 200g |
| 洋蔥切粒 | ¼個 | White onion, diced | ¼ |
| 蒜頭切碎 | 3瓣 | Garlic cloves, minced | 3 |
| 焗薯蒸熟後壓成薯蓉 | ½個 | Baking potato, steamed and mashed | ½ |
| 蔥切碎 | 2茶匙 | Scallion, chopped | 2 teaspoons |
| 蛋黃發開 | 1湯匙 | Egg yolk, beaten | 1 tablespoon |
| 日本麵包糠（如有需要） | 2湯匙 | Japanese panko (if needed) | 2 tablespoons |
| 橄欖油或米糠油 | 煎炸用 | Olive oil or rice bran oil | for frying |
| 已焗芝麻 | 1湯匙 | Toasted sesame seeds | 1 tablespoon |
| 蛋黃醬 | 用作醬汁 | Mayonnaise | for serving |

做法

製作馬介休前，先將鱈魚在室溫下解凍2小時。用鹽搓勻魚柳，用保鮮紙封好，放入雪櫃冷藏2小時。用餐巾紙印乾魚柳。將鱈魚放入蒸隔中。將水倒入平底鍋，水高5厘米，放入蒸隔和魚，蓋上蓋子然後蒸10分鐘，直至魚柳蒸熟為止。將鱈魚印乾，將魚柳剝成碎塊。小心那些細小的魚骨。

將碎魚塊、切碎的蝦、洋蔥、蒜頭、焗薯和蔥放入大碗裡，用手提式攪拌器以中速將材料攪拌10秒鐘。不要將材料過度攪碎，要保留蝦粒的口感。拌勻並試味。視乎鱈魚的鹹度，你可能不需要再作調味，如果鱈魚蝦蓉味道偏淡，便加入少許鹽。拌入蛋黃。若太濕而且散開不成形，便拌入麵包糠。

將1湯匙大小的鱈魚蝦蓉搓成球狀。重複製作約12個球，將它們放在一隻大碟上。封好並冷藏至少30分鐘。

將5cm油倒入中型平底鍋以大火加熱。分批將鱈魚蝦球放入油中煎炸，轉一至兩次，直至蝦球變成金黃色。不超過3分鐘。用罩籬將蝦球取出，放在餐巾紙上吸走多餘油份。趁蝦球還灼熱時灑上芝麻。

拌以蛋黃醬，可熱食或暖食。

How to cook

To prepare bacalhau, thaw the cod at room temperature. Rub all over with the salt, cover, and marinate in the fridge for 2 hours. Pat dry with paper towels. Place the cod in a steamer basket. Set the basket in a saucepan with 5cm of water, cover, and steam for 10 minutes, until cooked. Pat the cod dry and flake the fish into tiny shreds. Watch out for the tiny little bones.

In a large bowl, mix together the shredded cod, chopped shrimp, onion, garlic, potato, and scallion. Use a handheld blender to blend at medium speed for 10 seconds. Do not overmix— you want pieces of chopped shrimp to remain in the mixture. Mix well and taste. Depending on how salty the cod is, you may not need to season the mixture: If it tastes flat, add a healthy pinch of salt. Stir in the egg yolk. Add the panko if the mixture is too moist and doesn't hold together.

Pinch off about 1 tablespoon of the cod-shrimp mixture and roll into a ball. Repeat to make about 12 balls, setting them on a large plate. Cover and refrigerate for at least 30 minutes.

Heat 5cm oil in a medium saucepan over high heat. In batches, add the shrimp balls to the oil and fry, turning once or twice, until golden brown. This should take no more than 3 minutes. Remove with a skimmer and let drain on kitchen towels. Sprinkle with sesame seeds while they're still sizzling.

Serve warm or at room temperature, with mayonnaise.

# 新派焗蛋白布甸
## Neo Molotoff

Molotoff是一種鬆脆的蛋白餅甜點，由蛋白和糖製成，常見於葡萄牙和澳門。在我6歲到葡萄牙里斯本探親時，我第一次嚐到了Molotoff的味道。當時我不知道甜點的名字，我乾脆稱它為 "White Pudding"。Neo Molotoff是我7歲時的自創料理， 靈感來自著名Baked Alaska，一個由雪糕和蛋糕製成的甜點，周圍鋪滿蛋白脆餅，在焗爐焗至微微上色。Neo Molotoff是我做過最簡單的菜式，卻往往讓我的粉絲驚嘆不已。

我把Neo Molotoff作為全書最後一個食譜，作為我對所有粉絲的致敬。烹飪就是簡簡單單，充滿趣味、快樂。就像Neo Molotoff這樣簡單的菜式：不用小題大做，一切隨心。

Molotoff is an airy meringue dessert, made simply with egg whites and sugar, that is commonly found in Portugal and Macau. My first taste of Molotoff was six years old while visiting relatives in Lisbon, Portugal. Having no clue about the name of the dessert, I called it White Pudding. Neo Molotoff is my own take on the classic, inspired by the famous Baked Alaska, a dessert of ice cream and cake surrounded by meringue that is browned in the oven. This is also the easiest dish I've ever made, but something that will always "wow" my fans.

I am making this recipe the last in this book as a salute to all my fans. Cooking is all about fun and love. A simple dish like Neo Molotoff is the way to go: no fuss, all love.

## 材料（4人份量）

| | |
|---|---|
| 蛋白 | 2隻 |
| 糖 | 1湯匙 |
| 檸檬汁 | 1茶匙 |
| 蜜糖 | 2湯匙 |
| 薄荷葉切碎 | 2茶匙 |

## Ingredient (Makes 4 servings)

| | |
|---|---|
| Egg whites | 2 |
| Sugar | 1 tablespoon |
| Lemon juice | 1 teaspoon |
| Honey | 2 tablespoons |
| Mint leaves, chopped | 2 teaspoons |

## 做法

將焗爐預熱15分鐘至200°C。

在不銹鋼碗或普通碗中，以高速電動攪拌器將蛋白打發。逐少加入糖和檸檬汁，將蛋白打發至白色並挺身。整個過程不應超過4分鐘。

將蛋白刮出放入四個直徑10厘米的焗杯中。焗3分鐘左右，直至蛋白膨脹並略為著色。不要過度烘焗，這樣會令蛋白下塌。

於蛋白面放上蜂蜜和薄荷葉。剛出爐的Molotoff最美味，一定要立即嚐一口。

## How to cook

Preheat the oven at 200°C for 15 minutes.

Whip the egg whites in a stainless bowl or bowl with an electric mixer on high speed. Gradually add the sugar and lemon juice and whip until the whites are stiff. The whole process should take no more than 4 minutes.

Scrape the meringue into four 10cm ramekins. Bake for about 3 minutes, until the meringues expand and are lightly browned. Do not overbake as the meringue will collapse.

Top with honey and mint leaves. Molotoff is best served straight from oven and enjoyed immediately.

*Tips* 你也可以在這個蛋白甜點上放些芒果蓉或熱情果蓉。

You can also top the meringue with mango or passion fruit puree.

# 食譜 41
## Recipe 41

食譜41是什麼？如果你從頭到尾看過我的書，你應該已經掌握了如何烹調一些葡國菜。現在是時候製作自己的食譜了。邀請你加入我的美食之旅，分享烹飪的樂趣。把你的原創新式葡國食譜分享到我的臉書。我會在第二本烹飪書中介紹三款由讀者提供的食譜。好好享受烹飪之樂！

What is Recipe 41? It is the recipe created by you! If you have read my book from the beginning, you probably already have some ideas on how to cook Portuguese and Macanese dishes. Now it's the time to make your own. And time to join my Food Journey and share the fun of cooking. Post your original neo Portuguese/Macanese recipe on my page at. I will feature three reader recipes in my second book.

Enjoy and have fun!

 http://facebook.com/eufoodjourney

# 結語

2018年1月開始了這本書的製作，到了今天8月27日，這本書已經伴了我一段日子。寫下了這篇結語，也代表了一個新的里程碑，有不捨之情嗎？當然的啊！這8個月的創作過程中身邊也發生了很多的事情，爸爸進出了醫院幾次，在7月的時候也在醫院住了接近6星期，3名年長毛孩走了後我終於在今年5月再次收養了2名毛孩。我的Facebook page 追蹤人數終於過了10,000人！

身邊有幾位烹飪達人朋友相繼在這幾個月也出了食譜書，不知道他們有否想過我一直都有想的問題：What is next？我已經計劃好未來的方向，而未來的日子將會是更忙碌，更充實。真的沒有想過可以用自己的名字出版一本關於自己的書集，一切都是天時地利人和，竟然已經走出了一大步，我就應該再向前多走幾步。

這本書讓我認識到更多有關我父母的童年往事，也解答了很多我對爸爸和嫲嫲why this and why that 的問題。兩星期前，我有一位料理班同學給我看了一張照片，照片的人物是我的一家人和姑媽的一家人。姑媽在嫲嫲於1990年離開後也相繼離開了，而我們也慢慢的跟姑媽的家人失去聯絡。20年後，我們新的一代終於透過食物再次get connected。而我也從這位同學的口中知道了更多關於嫲嫲的story，而這些事又trigger off 我想問父母的一些問題。但看見爸爸現在還是在休養的階段，而媽媽又是開心的跟外孫忙著買新校服，我都不打算再問他們兩老了。

食物令我們走在一起，新的旅程也要展開了。

2018 年8月27日

# Epilogue

So I have done it! This is truly an amazing journey. And I still cannot believe I am sitting here and writing this epilogue. I have come so far and I can't and I won't remain where I am. Life is short, and there is still so much for me to learn and to give. Ham Har Chaan is my legacy, and I shall remember each and every one of you. You have all played a part in my food journey.

A new journey begins now.

August 27, 2018

特別鳴謝 WOLL 和 Epicurean 贊助廚具，
更要多謝 Ready To Cook 和 The Cooking Alley 借出場地供拍照之用。
Thank you WOLL and Epicurean for sponsoring all my cookwares,
and a big cheers to Ready To Cook and The Cooking Alley for giving us the space for our shooting.

# 鹹蝦燦之味
## The Ham Har Chaan Cookbook

Published in Hong Kong 2018
2018 香港出版

Copyright © John Rocha 2018
Photography copyright © John Rocha 2018

Author: John Rocha
Email: eurasianfoodjourney@gmail.com
Facebook: http://facebook.com/eufoodjourney
Instagram: eurasianfoodjourney
Whatsapp/tel: (852) 5503 8258

Design & Art Direction by Matthew Wu
Photography & Photo Direction by  Harry Chan ( IG: nth_hppns ) &
Ip Wai Lung ( IG: ip_wai_lung )

Published by Ada Wang
出版人：王凱思

WE Press Company Limited
香港人出版有限公司
14/F, Greatmany Centre 109-115 Queen's Road East Wan Chai, Hong Kong
香港灣仔皇后大道東109–115號智群商業中心14樓

Website: www.we-press.com   Email: info@we-press.com
Facebook: www.facebook.com/wepresshk
Instagram: we_press   Whatsapp/tel: (852) 6688 1422

ISBN: 978-988-13267-5-1

WE PRESS
香 港 人 出 版